Study Skills for Geography Students

Study Skills for Geography Students

A PRACTICAL GUIDE

PAULINE E. KNEALE

A Member of the Hodder Headline Group
LONDON • NEW YORK • SYDNEY • AUCKLAND

For Alistair and Gemma

First published in Great Britain in 1999 by
Arnold, a member of the Hodder Headline Group,
338 Euston Road, London NW1 3BH

http://www.arnoldpublishers.com

Co-published in the United States of America by
Oxford University Press Inc.,
198 Madison Avenue, New York, NY10016

British Library Cataloguing in Publication Data
A catalogue record for this book is available from the British Library

Library of Congress Cataloging-in-Publication Data
A catalog record for this book is available from the Library of Congress

ISBN 0 340 73172 9

1 2 3 4 5 6 7 8 9 10

Production Editor: Liz Gooster
Production Controller: Helen Whitehorn
Cover Design: Terry Griffiths

Typeset in Palatino and Gill Sans by Phoenix Photosetting, Chatham, Kent
Printed and bound in Great Britain by J.W. Arrowsmith Ltd, Bristol

CONTENTS

PREFACE

This book is for geography students who want information about research and study in Higher Education, designed to defuse confusion and build confidence. It assumes that most geographers would rather spend extra time in the field, tackling the issues involved in the geographies of consumers at hostelries, than in the library. Most people using self-help books or on training courses find that 90+ per cent of the material is already familiar, but the few new elements make it worthwhile. The 90 per cent increases your confidence that you are on the right track, and the remainder, hopefully, sparks some rethinking, reassessment and refining. The trick with university study is to find a combination of ideas that suit you, promote your research and learning, and add self-confidence. As with all texts, not all the answers are here, but who said study would be a doddle? This text is intended for reference throughout your degree, some items will seem irrelevant at first but be important later. There is a real difference between reading about a skill and applying the ideas in your degree. The **Try This** activities are designed to make the link between skills and their practical application, and to give you an opportunity to practise either mentally, or mentally and physically.

I hope you will enjoy some of the humour: this is not meant to be a solemn book, but it has serious points to make. The geographical jokes are dire. The four crosswords follow the style familiar to readers of UK broadsheet newspapers, two quick crosswords and two cryptic; all the answers have vaguely geographical connections. Keep in mind you should be enjoying studying geography at university, it is supposed to be exciting and it can be fun as well as a challenge.

KEEP SMILING!

ACKNOWLEDGEMENTS

Many thanks to all The University of Leeds geography undergraduates and postgraduates that survived the Study Skills module. A great many of the ideas you prompted are here, together with a pot-pourri of your examples and comments. Thanks also to the staff of the School of Geography, especially Mark Newcombe for the graphics, Debbie Phillips, Matt Stroh and Marcus Power. Members of The University of Leeds Staff and Student Development Unit and the Careers Service, Chris Butcher, Maggie Boyle, Val Butcher, Sue Hawksworth, Jane Conway. Professor George Brown, Nick Barstow, Sylvie Collins, Jenny Elm, Angela Oxley, Alan Richmond, Emily Saunders, Lynne Shaw, David Tucker and Chrissie Tunney who tried it out. Without TQA and Enterprise in Higher Education, all this would not be possible.

Thanks are due to the following for permission to reproduce diagrams and extracts: Arnold for extracts from *Progress in Human Geography* and *Progress in Physical Geography*; Carfax Publishing Limited, PO Box 25, Abingdon, Oxfordshire OX14 3UE, UK for material previously published in *Journal of Geography in Higher Education*. Victor Gollancz Ltd for permission to quote extracts from *Interesting Times* (1994) and *Maskerade* (1995) by Terry Pratchett. The Royal Geographical Society with the Institute of British Geographers for permission to use extracts from *Transactions of the Institute of British Geographers* and *Area*.

How does an Inuit planner build a new town?

Igloos it together.

1 WHERE DO YOU FIND STUDY SKILLS IN GEOGRAPHY?

'Teach? No,' said Granny. 'Ain't got the patience for teaching. But I might let you learn.'

(Pratchett, 1995, *Maskerade*)

This book discusses the skills needed to research and study for a geography degree. Some of the motivation for assembling it came from a student who said, 'The problem with first year was I didn't know what I didn't know, and even when I thought there was something I was supposed to know I didn't know what to do about it.' University can seem confusing, you are expected to learn independently, rather than being taught, but there is limited information about how to learn. This book might help. It is deliberately 'hands-on', making lots of practical '**Try This**' suggestions. It aims to add to your self-confidence in your research and study abilities, and save your time by acting as an ongoing resource. Rather than worrying about what will happen in a seminar, or how to do an on-line search, or reference an essay, look it up and carry on. You are already skilled in many areas like thinking, listening, note making and writing BUT reviewing your approach and refining your skills should prove beneficial.

The language and tone of the book are deliberately light-hearted, with some games for light relief. There are some terrible jokes, although being written down inevitably diminishes humour, keep smiling as you groan. Light relief is vital in study, if deep thinking leads to deep kipping, have a coffee, solve a crossword clue, but remember to go back to thinking after your break!

University is about taking personal control of what you do and how you do it. There is a departmental teaching agenda to follow to get your degree, and time to explore other avenues. If, in the process, these equip you for later life, that is a bonus.

A geography degree has two elements:

- **The geographical knowledge element,** including all the current theories from cultural and medical geography, the racing speed of a warm-based glacier, the consequences of deforestation, answers to fundamental economic problems like 'if money doesn't grow on trees, why do banks have branches?', the geo-political implications of destabilisation in the Balkans. The scope is global because, of course, 'geographers do it world-wide'.
- **The skills element.** Often called transferable skills, they allow geographers to be efficient researchers and have a longer-term benefit in the work place. Most geography graduates will acquire practical experience of the skills and attributes shown in Figure 1.1. Some will be picked up by osmosis, others will be taught at varying levels of detail.

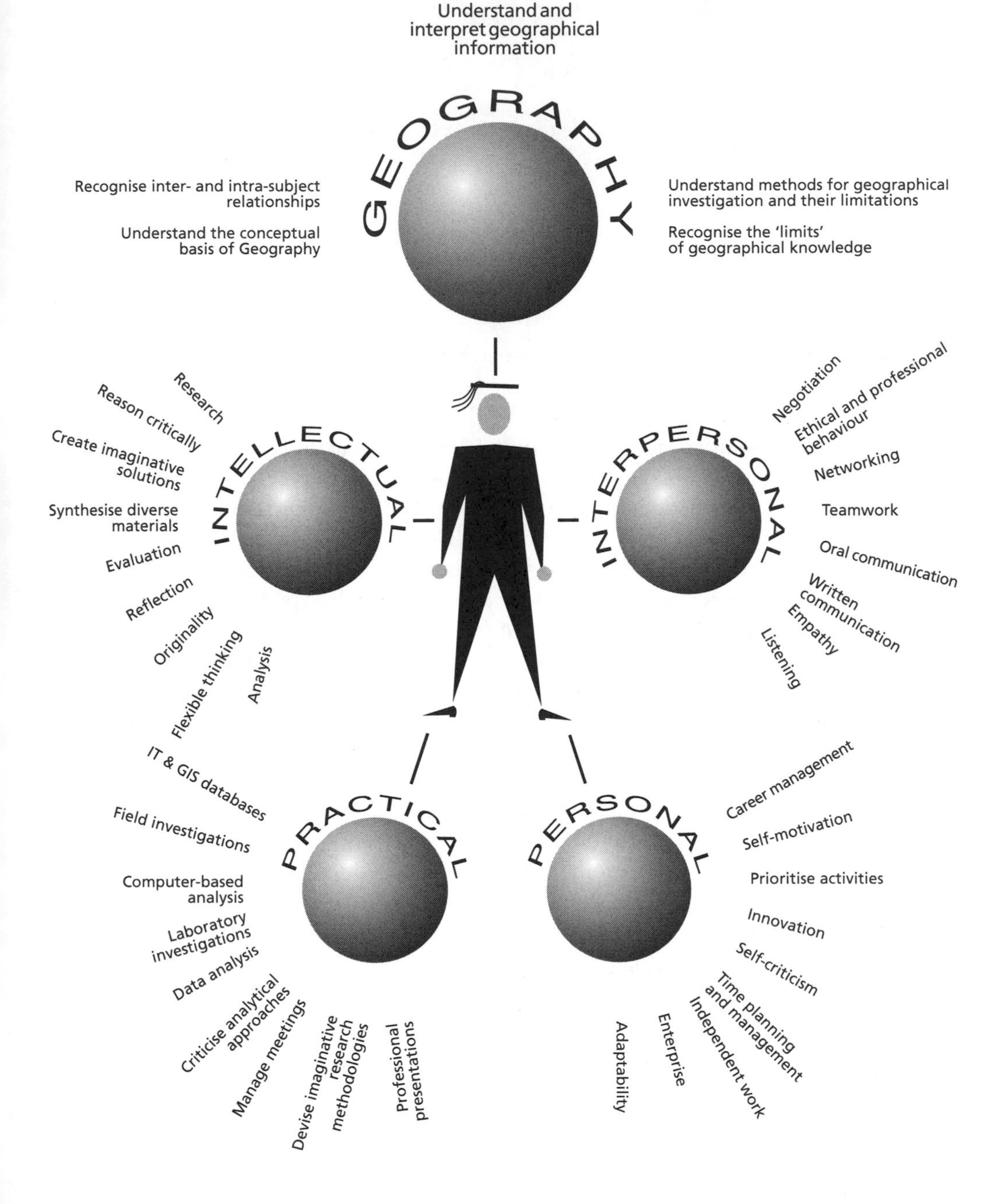

Figure 1:1 Skills and attributes of a geography graduate

In the last years of the twentieth century, UK student numbers have expanded and the emphasis has switched from lecturers teaching to students learning. Self-motivated learning is vital in life, enabling you to keep abreast of developments and initiatives. Employment is unpredictable. Job market and company requirements change rapidly. Employers need individuals who are flexible about their careers. An efficient graduate is someone who sees their career as a process of work and learning, mixing them to extend skills and experience. This is the essence of lifelong learning.

In the jargon of career management and personal development, the phrase 'transferable skills' is readily quoted. To add value to your degree, you need to recognise and reflect on what you do every day in your course (see Chapter 2), and understand where these skills have market value. Employers claim to be happy with the academic skills students acquire, such as researching, collating and synthesising new material, but also want graduates with skills like listening, negotiating and presenting. Any strengthening of your skills and experience of skill-based activities should add to your self-confidence and improve your performance as a geographer and as a potential employee.

In addition to traditional geography skills, your degree will give you the opportunity to experience some of the latest electronic trends, tele-working, surfing the world wide web (www), electronic journals, video-conferencing, e-mail, bulletin-boards, databases and spreadsheets, and video. University encourages you to get wired, get trained and build your own electronic resource base. The technology may seem daunting but it is fun too. (And if some five-year-old proto-anorak wearer can manage, so can you!)

The importance of graduates acquiring skills as well as knowledge was reinforced by the Dearing Report (NCIHE, 1997) which defined four key graduate skills, and happily, Geography degrees are awash with them (see Table 1.1). However, these four need some unpacking to show what is involved. AGCAS (1997) lists the self-reliance skills that organisations and companies desire, most elements involve Communication, IT and Learning how to Learn:

- communication skills, both written and oral, and the ability to listen to others;
- interpersonal or social skills, the capacity to establish good, professional working relationships with clients and colleagues;
- organisational skills, planning ahead, meeting deadlines, managing yourself and co-ordinating others;
- problem analysis and solution, the ability to identify key issues, reconcile conflicts, devise workable solutions, be clear and logical in thinking, prioritise and work under pressure;
- intellect, judged by how effectively you translate your ideas into action;
- leadership, many graduates eventually reach senior positions managing and leading people;

Graduate Skills	In Geography Degrees
Communication	All modules. Oral and written communication in seminars, tutorials, workshops, debates, group work, practicals, field class and all assessments.
Numeracy	All modules involving statistics, computing, databases, programming, data handling, calculations in practicals, field exercises, projects and dissertations.
Use of Information Technology	Most modules. On-line research activities. Word processing. Modules using graphics, statistics, databases, programming and GIS (geographical information systems).
Learning how to Learn	All modules. Taking personal responsibility for learning as an individual, and in group research, fieldwork, project and dissertation.

Table 1:1 Where to find NCIHE (1997) skills in action

- teamwork, working effectively in formal and informal teams;
- adaptability, being able to initiate and respond to changing circumstances, and to continue to develop one's knowledge, interests and attitudes to adapt to changing demands;
- technical capability, the capacity to acquire appropriate technical skills including scheduling, IT, statistics, computing, data analysis and to update these as appropriate;

IT skills disappear without practise

- achievement, the ability to set and achieve goals for oneself and for others, to keep an organisation developing.

By graduation you should feel confident in listing these skills on a curriculum vitae (CV), and be able to explain where in the degree these abilities were practised and demonstrated.

1.1 Learning activities at university: what to expect, and spot the skills!

Geography degrees are traditionally divided into three years called either Years 1, 2 and 3, or Levels 1–3. An additional year may be intercalated for an industrial placement or a year abroad. A year is, typically, divided into 10 or 12 teaching blocks called modules or units, addressing geographical or related topics. Geography degrees are usually progressive, which means that the standards and difficulty increase each year, and modules in later years build on experience and learning in earlier years. This section outlines the main activities at university and some of the skills that can be practised during them.

Lectures

Believing any of the following statements will seriously damage your learning from lectures:

- In good lectures the lecturer speaks, the audience takes very rapid notes and silence reigns.
- The success of a lecture is all down to the lecturer.
- A great lecturer speaks slowly so students can take beautifully written, verbatim notes.
- Everything you need to know to get a first class degree will be mentioned in a lecture.
- Lectures are attended by students who work alone.

Lectures are the traditional teaching method, usually about 50 minutes long, with one lecturer and loads of students. If your lectures involve 300+ students they may seem impersonal and asking questions is difficult. Good wheezes to manage lectures include:

☺ Get there early, find a seat where you can see and hear.

☺ Have a supply of paper, pens and pencils ready.

☺ Get your brain in gear by thinking 'I know I will enjoy this lecture, it will be good, I really want to know about…'; 'Last week we discussed…, now I want to find out about…'!!

☺ Before the lecture, read the notes from the last session, and maybe some library material too. Even 5–10 minutes worth will get the old brain in gear.

☺ Look at handouts carefully. Many lecturers give summary sheets with lecture outlines, main points, diagrams and reading. Use these to plan reading, revision and preparation for the next session.

- ☺ Think critically about the material presented.
- ☺ Revise and summarise notes soon after a lecture, it will help in recalling material later. Decide what follow up reading is required.
- ☺ Ask questions of yourself and others.

Skills acquired during lectures include understanding geographical issues, recognising research frontiers and subject limitations.

Tutorials: What are they? What do you do?

Usually tutorials are a 50-minute discussion meeting with an academic or post-graduate chairperson and 4–8 students. The style of tutorials varies between departments, but there is normally a set topic involving preparation. You might be asked to prepare a short talk, write an essay, write outline essays, prepare material for a debate, review a paper, produce a short computer program, and to share your information with the group. The aim is to discuss and evaluate issues in a group that is small enough for everyone to take part. Other jolly tutorial activities include brainstorming examination answers, comparing note styles, creating a poster, planning a research strategy, discussing the practicalities of a fieldwork proposal, evaluating dissertation possibilities, and the list goes on.

The tutor's role is NOT to talk all the time, is NOT to teach, and is NOT to dominate the discussion. A good tutor will set the topic and style for the session well in advance, so everyone knows what they are doing and will let a discussion flow, watch the time, make sure everyone gets a fair share of the conversation, assist when the group is stuck, and sum up if there is no summariser to do so as part of the assignment. A good tutor will comment on your activities, but tutorials are YOUR time.

Some tutors will ask you to run a couple of tutorials in their absence, and to report a summary of the outcomes. This is not because tutors are lazy, but because generating independence is an important part of university training. Student-led and student-managed tutorials demonstrate skills in the management of group and personal work. When a tutor is ill, continuing unsupervised uses the time effectively. A tutor may assign, or ask for volunteer chairpersons, timekeepers and reporters to manage and document discussions.

Your role is to arrive at tutorials fully prepared to discuss the topic NO MATTER HOW UNINTERESTED YOU ARE. Use tutorials to develop listening and discussion skills, to become familiar with talking around geographical issues and build up experience of arguing about ideas.

Top Tips

- ✔ Taking time to prepare for tutorials will stop (or reduce) nerves, and you will learn more by understanding a little about the topic in advance.
- ✔ Reviewing notes and reading related material will increase your confidence in discussions.
- ✔ Asking questions is a good way of saying something without having to know the answer.
- ✔ Have a couple of questions or points prepared in advance, and use them early on, get stuck in.
- ✔ Taking notes in tutorials is vital. Other people's views, especially when different from your own, broaden your ideas about a topic, but they are impossible to recall later unless noted at the time. Tutorial notes make good revision material.

Tutorial skills include listening, communication, presentation, critical reasoning, analysis, synthesis, networking and negotiation.

Seminars

Seminars are a slightly more formal version of a tutorial, with 8–25 people. One or more people make a presentation for about half the allotted period, leaving ample time for group discussion. Seminars are a great opportunity to brainstorm and note the ideas and attitudes of colleagues. Spot extra examples and approaches to consider later. Take notes.

Even if you are not the main speaker, you need to prepare in advance. In the week you speak, you will be enormously grateful to everyone who contributes to the discussion. To benefit from this kind of co-operation you need to prepare and contribute in the weeks when you are not the main presenter.

ONLY TO BE READ BY THE NERVOUS. (Thank you.) If you are worried, nervous or just plain terrified, then volunteer to do an early seminar. It gets it out of the way before someone else does something brilliant (well, moderately reasonable) and upsets you! Acquiring and strengthening skills builds your confidence so seminars appear less of a nightmare. By the third week you will know people and be less worried.

Seminar skills include discussion, listening, analysis, teamwork, giving a professional performance and networking more connections than BT.

Workshops or large group tutorials

Workshops are classes with 12–35 students, that support lecture and practical modules. They have very varied formats. They may be called large group tutorials, support classes or revision classes. There will usually be preparation work and a group activity. Tutors act as facilitators, not as teachers. Expect a tutor to break large groups into sub-groups for brainstorming and discussion. These sessions present a great opportunity to widen your circle of friends and to find colleagues with similar and diverse views.

Workshop skills are the same as for tutorials and seminars with wider networking and listening opportunities.

Computer and laboratory practicals

Practical classes and fieldwork are the 'hands-on' skills element of a geography degree. Many departments assign practical class time, when tutor support is available, BUT completing exercises and developing your proficiency in IT, computing, and laboratory skills will take additional time. Check the opening hours of computer laboratories on campus.

In laboratory practicals always take note of safety advice, wear lab coats and safety glasses as advised, and please don't mix acids without supervision. Most laboratory staff are trained in first aid, but would rather not have to practice on you.

Assessment

Assessment comes like Christmas presents, regularly and in all sorts of shapes. Assessments should be regarded as helpful because they develop your understanding. There are two forms:

- ✔ Within module assessment of progress, where the marks do not count (formative), and usually involve some feedback.
- ✔ Assessments where the marks do count (summative). Feedback may or may not happen, depending on the test and system. The results eventually appear on your degree result notification for the edification of your first employer who wants written confirmation of your university prowess.

There is a slight tendency for the average student to pay less attention to formative, within module assessments, where the marks do not count. Staff design formative tests because they know 99 per cent (± 1 per cent) of students need an opportunity to 'have a go', to get an insight into procedures and expected standards, when marks are not an issue.

Assessments are very varied: examinations, essays, oral presentations, seminars, posters, discussion contributions, debates, reports, reviews of books and papers, project designs, critical learning log, field performance, laboratory competence, computer-based practicals, multiple choice tests ... It is a matter of time management to organise your life and them (Chapter 3). You should know in advance exactly how each module is assessed and what each element is worth. Many modules have mixed assessments, so those who do very well in examinations or essay writing are not consistently advantaged. Many departments have standard assessment criteria. Get hold of your own departmental versions or see examples for essays (Figure 12.2), oral presentations (Table 10.1), practical reports (Figure 15.2), dissertations (Figure 18.2) and poster presentations (Figure 23.3). If you cover all the criteria then the marks come rolling in.

Amongst the many skills enhanced by assessments are thinking, synthesis, evaluation, originality and communication.

Non-academic learning

Do not underestimate what you know! In your years at university you also acquire loads of personal skills, like negotiating with landlords, debt crisis management, charming bank staff, juggling time to keep a term-time job and deliver essays on time, being flexible over who does the washing up and handling flat mates and tutors.

1.2 The research process

All students are 'reading' for a degree, finding out for themselves, the research process. Taught activities involve 20–50 per cent of the timetable, leaving 80–50 per cent for personal research into sub-topics, and study activities which reinforce understanding. The scope of geography is vast, certainly global and at times interplanetary. Covering all aspects is impossible. Every geographer is aiming to maximise their research activities, to expand their geographical knowledge, within the constraints of time, facilities and energy. Consequently, working in a group (see Chapter 13) can be seriously beneficial.

Geographical issues are not simple. Lecturers will indicate what is already well understood and discuss areas of the subject where we are less sure about what happens, pointing out where knowledge is missing, provisional, uncertain and worthy of further investigation. Most geographical issues are highly interlocking and multidimensional. Geographers' greatest strengths are their ability to take complex, unclear and at times contradictory, information from a wide range of sources and synthesise it to make sense of the picture at a range of scales.

By the time you graduate you should have an enhanced ability to recognise both the boundaries of knowledge of geography, what is known and what is not known, and what you as an individual know and do not know. Recognising the boundaries of one's own expertise is a relevant life skill. Someone who does not understand the implications of their actions in changing procedures, for example, is a potential danger to themselves and the wider community. University learning is not about recalling a full set of lecture notes. It is about understanding issues and being able to relate and apply them in different contexts.

Recognising the links between geography and other disciplines will enhance your research ability. Increasing numbers of geography students have no formal school geography background. If you are one of these students, remember this is not a disadvantage, you are obviously interested in geography, and all topics are fresh and not confused by half-remembered notes (see p. 243). Geography draws on many subjects for theory and insights, it is likely that several of your lecturers did not do geography at university. Use the background information and skills you have from other subjects to strengthen your geographical research. Mature students, with longer experience of life, politics, social conditions and general knowledge have an extensive skill base to build on.

1.3 How to use this book

No single idea is going to make a magic difference to your learning, but taking time to think about the way you approach tasks like reading and thinking, listening and writing, researching and presenting should help your efficiency rate. Studying is a personal activity. There are no 'right' ways, but there are tips, techniques, short cuts and long cuts.

If you attempt to read the whole book in one go you are likely to feel got at and preached at. *This is not the idea.* Look through the chapter headings and index. If you are concerned or stuck then, hopefully, there is a useful section. Use it as a handbook throughout your degree. Some parts are relevant for level 1, others like the dissertation advice (Chapter 18), will matter more in the last year. When you have an essay or book review to write, look at the relevant chapters. No one expects you to know the whole of the Greek alphabet or all the Latin names for plants, but if someone mentions either there is a bit of this book (sections 24.9 and 24.10) that can act as a reference. The key is to view it as encouragement. Build on your existing skills and follow the skills route (see Figure 1.2). There are lots of **Try This** opportunities suggesting ways of practising or applying your skills. Adapt them to your needs. Treat them as part of your geography learning rather than as an isolated skill exercise. Experience is built on doing things, not on watching them.

The aim of the book is to encourage and build your self-confidence in your skills, and to show where they are transferable and marketable. Your task is to decide to what extent you agree with the ideas in the book, and can apply them for your own purposes. In some ways this is like a cookery book, many of the **Try This** activities are recipes, BUT adapt, garnish, modify and extend them. Be inventive with the ingredients! University research is a creative activity, like cooking. Some statements are deliberately controversial, designed to encourage thinking. Most of the figures and examples are deliberately 'less than perfect'. You are asked to consider how they can be improved, and what needs changing, as a way into active criticism. Universities provide IT facilities and gymnasiums, but getting more skilled means 'working out'. The first year is a good time to practise and enhance study skills, but it is important to keep practising and reflecting throughout your degree.

START FIRST YEAR
(Unaware of the rules?)

AWARENESS
(Could be better?)

LOADS OF PRACTICE

FIELD AND PRACTICAL ACTIVITIES

Improvement

reinforcement

INFORMED

ability
(Trained)

CONFIDENT
CAREER
Application

encouragement encouragement encouragement

Figure 1:2 The skills route

1.4 References and Further Reading

There are a tremendous number of generic skills texts. For a background on what geography involves see:

Rogers, A., Viles, H. and Goudie, A. 1992 *The Student's Companion to Geography*, Blackwell, Oxford.

For skills, see Chapter 26 and:

AGCAS 1997 *Making Applications*, Graduate Careers Information, Association of Graduate Careers Advisory Services, Manchester.

Bucknall, K. 1996 *Studying at University: how to make a success of your academic course*, How To Books, Plymouth.

Buzan, T. 1995 *Use Your Head*, (Rev. edn), BBC Books, London.

Drew, S. and Bingham, R. (eds) 1997 *The Student Skills Guide*, Gower, Aldershot.

NCIHE 1997 *Higher Education in the Learning Society*, Report of the National Committee of Inquiry into Higher Education, HMSO, Norwich, [on-line] http://www.leeds.ac.uk/educol/ncihe Accessed 10 January 1999.

Northedge, A. 1990 *The Good Study Guide*, The Open University, Milton Keynes.

Northedge, A., Thomas, J., Lane, A. and Peasgood, A. 1997 *The Sciences Good Study Guide*, The Open University, Milton Keynes.

Pratchett, H.T. 1995 *Maskerade*. Victor Gollancz, London.

Geograms 1

Try these geographical anagrams. Answers on p. 252.

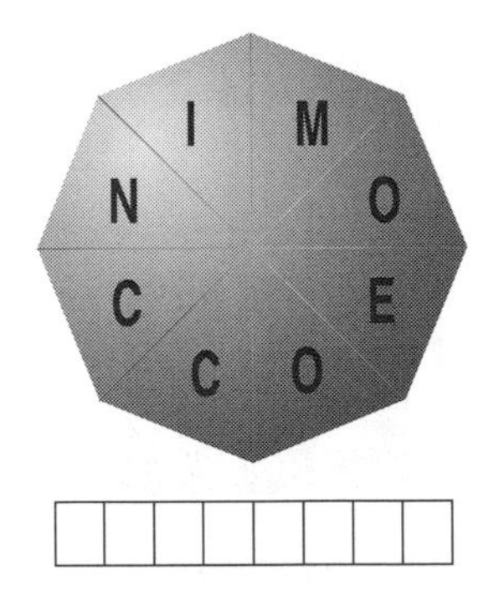

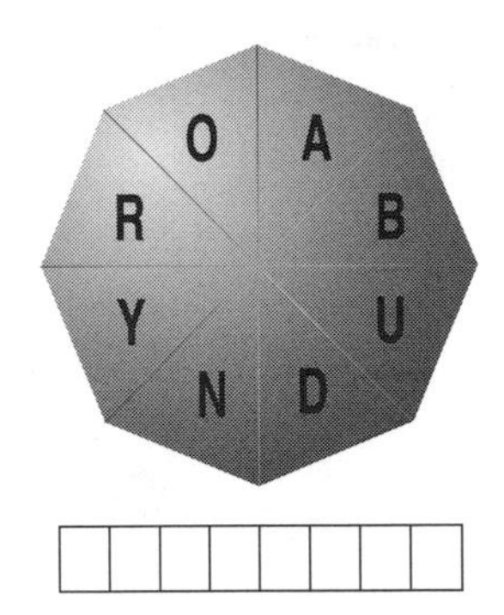

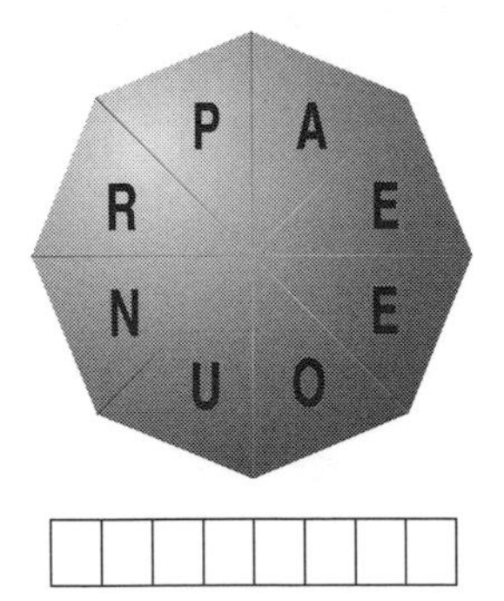

2 REFLECTION SKILLS, REVIEWING AND EVALUATING, ADDING VALUE TO YOUR DEGREE

Always remember you are unique, just like everyone else.

Geography students can be described as autonomous lifelong learners taking responsibility for planning their own learning. This is a complicated way of saying it is up to you to decide what to do and when to do it. Degrees involve personal decisions about what to read, research, ignore, practise, panic over, ... Reviewing reduces associated worries and provides some rational options when there is a choice, although many decisions are motivated by a looming submission deadline. Taking control and responsibility for your work can seem scary, and the lack of guidance about academic work is one reason why many new university students feel disoriented, chucked in the deep end without a Baywatch lifeguard in sight.

This chapter discusses why evaluation and reflection are useful skills, and how reflection techniques can give you a framework to help your decision-making in your studies. There are a variety of **Try This** activities because everyone has their own needs and priorities, and different activities will be relevant at different times of year and in different years of your degree. Most activities will benefit from a 'mental and physical (pen in hand)' approach. The knack is to develop this style of thinking so that it becomes automatic; reviewing becomes a process you can do while warming-up at kick boxing or cleaning out the parrot cage.

The skill benefits from this chapter are thinking, evaluation, reviewing and reflection. Reflection skills are not easily taught, or acquired overnight. They develop with experience and maturity. They are the outcome of thinking actively about experiences and placing them in a personal context, an iterative process.

2.1 WHY REFLECT?

Sometimes you have to step back to leap forward.

Adding value to your degree

Some students find it hard to see how parts of their degree course inter-connect, which is demotivating. One quote from a final year student encapsulates this:

For most of my degree we did lots of different modules and they were interesting in a way, but I couldn't see why we were doing economic geography and then statistics and the resources bits and so on. I couldn't really see the point at working at the bits. It was about half way through final year when all the bits began to make sense and slot together. And then it got to be really interesting and I could see why I should be doing more reading and I did quite well in the last semester.

Taking a little time to think about inter-connections between modules, 'why economic geography principles are important for cultural geographers or flood forecasters', or 'where statistical tests can be used in your dissertation', can give your modules more cohesion and be motivating.

Increasing employability

Many employers look for enthusiastic graduates with skills of articulation and reflection, those who can explain, with examples, and evaluate their experience and qualities. Recruiters want to identify people with the awareness and self-motivation to be pro-active about their learning. The ability to teach oneself, to be aware of the need to update one's personal and professional expertise, and to retrain, is vital for effective company or organisational performance and competition (Boud *et al.* 1993; Harvey *et al.* 1997; Hawkins and Winter 1995).

Your academic background may be of little interest to an employer. Whether you are expert in modelling the spread of disease, have researched Mongolian housing patterns or abseiled down a glacier is not important. What is relevant is that faced with the task of researching the market for a new type of fluffy bunny you can apply the associated skills and experience gained through researching the nineteenth-century development of the Co-operative Movement in Rochdale, (thinking, reading, researching, presentation, making connections) to the marketing of Peter Rabbit clones. It is your ability to apply the skills acquired through school and university in a workplace role that an employer will value.

Remember that an employer is looking for a mix of skills, evidence of your intellectual, operational or practical and interpersonal skills (look back at Figure 1.1). Your geographical intellectual skills are demonstrated by your degree certificate. You need to include your practical, transferable and 'hands-on' skills on your CV. Keeping a record of your thoughts on forms like those in the **Try This** activities here or in a diary or journal style log will pay off when filling in application forms. They will remind you of what you did and of the skills involved.

If you plan to work in very large multinational companies where skills training are a big budget item and there are many training courses, your lifelong learning will be enhanced by company policy. If you want to establish your own business, or to join small and medium-sized enterprises (SMEs) with small numbers of employees and budgets, then your university skills will be directly beneficial. Geographers with 'skills' are valuable employees.

2.2 Getting started

Some businesses require graduate trainees to keep a daily log in their early years of employment. It encourages staff to assess the relative importance of tasks and to be efficient managers of their time. It is a reflective exercise where at 4.50 p.m. each day, you complete a statement like:

> I have contributed to the organisation's success/profits today by ....
>
> ....
>
> I was fully skilled to do ....
>
> I was less capable at ....
>
> Other comments ....

At the end of the week or month, these statements are used to prioritise business planning and one's CPD (Continuing Professional Development). It is an activity that most new employees hate. However, most will admit, later, that it taught them an enormous amount about their time and personal management style, and wished they had started sooner. In time, this type of structured self-reflection becomes automatic, individuals continually evaluate their personal performance and respond accordingly.

The geography student equivalent is:

> I contributed to my geography degree today by ....
>
> ....
>
> I could have been more efficient at doing this if ....
>
> Tomorrow I am going to ....

Reviewing the whole day may seem too much of a drag. It may be easier to start by responding to statements like:

> What I have learned from this paper/lecture is ....
>
> It agrees with ....
>
> It has a methodology we could use for ....
>
> In future I will ....

There are six **Try This** activities in this chapter, each suited to different stages on a degree course. These kinds of forms can be used to build up a learning log, recording your university experience. A learning log can act just as a diary, somewhere to note activities and skills, but a reflective log asks for a more detailed, reasoned response.

2.3 Reflecting on your degree experience

Try This 2.1 asks you to articulate your feelings about your current, personal approach to learning and your degree course. If you then evaluate your response and decide to do something in response to your reflection, you are taking charge.

TRY THIS 2.1 – Personal reflection

This is a random list of degree skills with some examples of student comments. For those relevant to you at present, make a self-assessment of your current position and a reflective explanatory comment. Expand the list to suit your own position.

Skill	*Current skill level* *High*	*Low*	*Comment*
Delivering essays on time.			
Speaking in tutorials and seminars.		*	*I must say things in tutorial. I know what I want to say but it all seems so obvious I feel silly, so I guess I need to get stuck in.*
Listening carefully to discussions and responding.			
Being efficient in library research.			
Knowing when to stop reading.		*	*I read to the last minute, and then rush the writing. Some reading deadlines would help.*
Making notes.			
Using standard grammar, spelling and punctuation in writing.			
Using diagrams to illustrate points.			
Organising ideas coherently.	*		*I can do this, but include all ideas.*
Including relevant information in essays.		*	*I try to include everything in an essay, to show I have done some reading!*

Skill	*Current skill level* *High Low*	*Comment*
Including accurate information in reports and essays.		
Drawing the threads of an argument together and so developing a logical conclusion.		
Summarising information from different sources.		
Putting ideas into my own words.		
Negotiating.		
Disagreeing in discussion without causing upset or being upset.		
Using databases to extract geographical information.		
Being more open to new ideas.	*	*Might buy a paper. I could take notes in tutorials. Reading might be OK!*
Identifying the important points.	*	*Underlining or highlighting bits in notes would help and not take as long as re-writing things.*
Being organised and systematic.	*	*I must sort out my room, buy some files. If I used a diary it would help.*
Trying out new ideas.	*	*Listen to other people more. Take more time to look at options, I tend to make decisions without really thinking.*
Making time to sit down and think about different ideas.	*	*This seems really odd, because you sort of do thinking all the time. It's not really a cool activity. Could try when no one knows – in bed maybe.*
Using IT for word processing.		
Reading appropriate material.		

Try This 2.1 is a self-assessment exercise that you might want to repeat after a term or semester. Please note that when people self-assess a skill before and after an activity, the assessment at the end is frequently lower than that at the start. Although the skill has been used and improved during the activity, by the end, it is possible to see how further practise and experience will lead to a higher skill or competence level. Now look at **Try This 2.2**. In thinking about your strengths and weaknesses talk to family and friends, ask what skills you have and what you do well.

TRY THIS 2.2 – Being pro-active about skill development

Having completed **Try This 2.1** go back to the list and highlight three skills you would like to be pro-active about in the next three months. Now make some notes about how to be active about these three issues. Like New Year resolutions, this activity can bear re-visiting.

- Reflection is reinforced when you write down your thoughts or speak them aloud.

2.4 Within module reflection

Some modules will have a learning log as part of the assessment process. On fieldclass or as part of a laboratory or dissertation it serves the dual purpose of a diary and a reflective statement. Practise by answering some of the questions in **Try This 2.3**. With a little thought you can adapt most of them for most modules, or for your degree as a whole.

TRY THIS 2.3 – Reflecting on a class or module

Pick a module, or part of it, and answer the questions that are appropriate. Some sample answers from geography students are included on p.252.

- What I want to get out of attending this module is..
- In what ways has this module / session helped you to develop a clearer idea of yourself, your strengths and weaknesses?
- Record your current thinking about the skills you *can* acquire from this module. Tick those you want to develop, and make a plan.
- I have discovered the following about myself with respect to: decision-making . . .; research . . .; thinking . . . .
- How efficiently did you (your group) work?
- What skills did you use well, what skills did other members of the group use well?
- How did preparation for your (the group) presentation progress? What were your concerns?

- What skills were lacking in you (the group) and caused things to go badly?
- How did you (your group) make decisions?
- What have you learned about interview technique? / asking questions? / planning laboratory work? / investigating in the field?
- What did you enjoy most about the exercise / session / module / degree course?
- What did you enjoy least about the exercise/ session / module / degree course?
- What was the biggest challenge to you in this exercise/ session / module / geography degree?
- Give personal examples that illustrate two of the skills and attributes you have gained from this module.

The last four questions in **Try This 2.3** are frequently asked in interviews. Thinking of an answer in advance gives you more chance to enthuse and be positive. You may not have had much experience of some skills, but any experience is better than none.

TRY THIS 2.4 – Reflecting on a day

Brainstorm a list of things that happened (5 minutes maximum, just a back-of-an-envelope list) e.g.:

Went shopping.
Had hair cut.
Went to Dr Impossible's lecture.
Talked to Andy.

Then brainstorm a list of things that made the class/day unsatisfactory, e.g.:

Andy talked to me for hours.
Bus was late.
I didn't understand what Dr Impossible was going on about.
Printer queues were hours long.

Leave the two lists on one side for a couple of hours. Then grab a cup of coffee, a pen, re-read your lists and make a note about where you might have saved time, or done something differently. Consider what might make life more satisfactory if these situations happen again:

Natter to Andy for an hour MAXIMUM! over coffee and leave.
Take a book on the bus.
Have a look at Dr Impossible's last three lectures. If it still doesn't make sense I will ask my tutor or Dr Impossible.
Need to take something to read, or do on-line www searches, while printouts are chugging through.

Sometimes one recognises that a particular lecture, laboratory class or day has passed without being of any real benefit to one's degree! Try to identify why, **Try This 2.4** and **Try This 2.5** contain some ideas.

TRY THIS 2.5 – Reflecting on a class

The seminar was: very useful / useful / OK / boring.

The good things about it were..

The things that hindered it were...

To make it useful in retrospect I will .... (read ...., check out .... on the www, talk to ....)..

In order to get more from the next seminar I will..

By analysing and reflecting on what is happening in a structured manner, you will feel more in control. Following up one in ten of the ideas you have will be an improvement on the present. It is possible to review in your head, the bus stop might be a good location, but writing down your list is likely to be more beneficial. It will develop your reflective skills, and prompt action on those reflections if you put the lists somewhere where you will see them again.

2.5 What to do first?

There are many competing demands on your time, and it is not always obvious whether the next research activity involves finishing a practical report, browsing the library shelf for next week's essay or reading another paper. Reflect on who or what takes most of your time. Some tasks do take longer than others, but the proportions should be roughly right. Questions which encourage prioritising tasks include:

- Why am I doing this now, is it an urgent task (see also Table 3.1)?
- Is the time allocated to a task matched by the reward? For example, it is worth considering whether a module essay worth 50 per cent deserves five times the time devoted to a GIS practical worth 10 per cent.
- When and where do I work best? Am I taking advantage of times when my brain is in gear?
- How long have I spent on this web search, seminar preparation, mapping practical, Africa essay? Were these times in proportion? Which elements deserve more time?
- Am I being interrupted when I am working? If I worked somewhere else would that help?

- Who causes me to take time out? Are there ways of limiting this by say an hour a week?

2.6 Start on your CV now

Use reflective material to amplify your curriculum vitae (CV). The thought of leaving university, applying for jobs, and starting a career is probably as far from your mind as the state of the Bolivian economy. Nevertheless, for those desperate for money and applying for summer jobs, having a focused CV could significantly increase the chance of selection for that highly paid shelf-filler or burger-bar job. Building up a CV as your degree course progresses can save time in the last year. An electronic CV designer may be available on-line, the university careers service should be able to advise you. Reflecting on your skills at an early stage may highlight the absence of a particular 'skill'. There is time to get involved in something which will demonstrate you possess that skill, before the end of your degree. Have a go at **Try This 2.6**.

TRY THIS 2.6 – Skills from my geography degree

Expand and tailor this list for your degree, from your university. Be explicit in articulating the skills and the evidence. Update it each semester. There are a few starter suggestions in the second column.

Skills acquired from MY degree to date	
Numeracy	*Statistics modules in years 1 and 2. Calculations for science laboratory experiment. I completed a financial balance sheet for a set of laboratory experiments and for my dissertation.*
Able to meet deadlines-essays, reports, practical write-ups, etc.	*All essays completed in time. Organised group project and planned the mini-deadlines that kept us on track.*
Team work skills-workshop group work	
Communication and presentation skills, tutorial, seminars and presentations	
Computing skills	*Validation and calibration of a simulation model in computer practicals.*

TRY THIS 2.6 – *Continued*

Skills acquired from MY degree to date – *continued*	
IT skills	*Word processing of essays and 8000 word dissertation. Included Excel diagrams, PowerPoint graphics and output from SPSS packages.*
Able to put ideas across	
Able to work individually	
Time management skills	
Organisational skills	*Final year dissertation, organised personal fieldwork in nature reserve, required co-ordination with landowners, warden and with laboratory staff for analytical facilities.*
Self-motivated	
Able to prioritise tasks	
Problem-solving	

If you have forgotten what skills your modules involved, look back at the course outline. It is likely to include a statement like: 'On completion of the module students will have ...'. Use this kind of statement to amplify your CV and jog your memory.

Try This 2.6 does not include skills acquired through leisure pursuits or work experience. Compile a second list from those experiences. Driving, shorthand, stocktaking, flying, language skills, writing for newspaper or magazine, treasurer, secretary and chair of societies ... involve skills like time management, negotiation, listening, writing reports. Work experience does not have to be paid work, voluntary activities can give you valuable experience that pays dividends on a CV.

2.7 References and Further Reading

On the changing nature of work and the importance of skills, see the following:

Boud, D., Cohen, R. and Walker, D. 1993 *Using Experience for Learning*, Society for Research in Higher Education and Open University Press, Buckingham.

Harvey, L., Moon, S., Geall, V. and Bower, R. 1997 *Graduates' Work: Organisational Change and Students' Attributes*, Centre for Research into Quality and Association of Graduate Recruiters, Birmingham.

Hawkins, P. and Winter, J. 1995 *Skills for Graduates in the 21st Century*, The Association of Graduate Recruiters, Birmingham.

Purcell, K. and Pitcher, J. 1996 *Great Expectations; the new diversity of graduate skills and aspirations*, Higher Education Careers Services Unit, Careers Service Trust, Institute for Employment Research, Manchester.

If these are unavailable in your library or careers service offices do a library keyword (see p.35) search using career, lifelong learning, graduate skills and career development.

Wordsearch 1

Find 48 geographical terms. Answers on p.253.

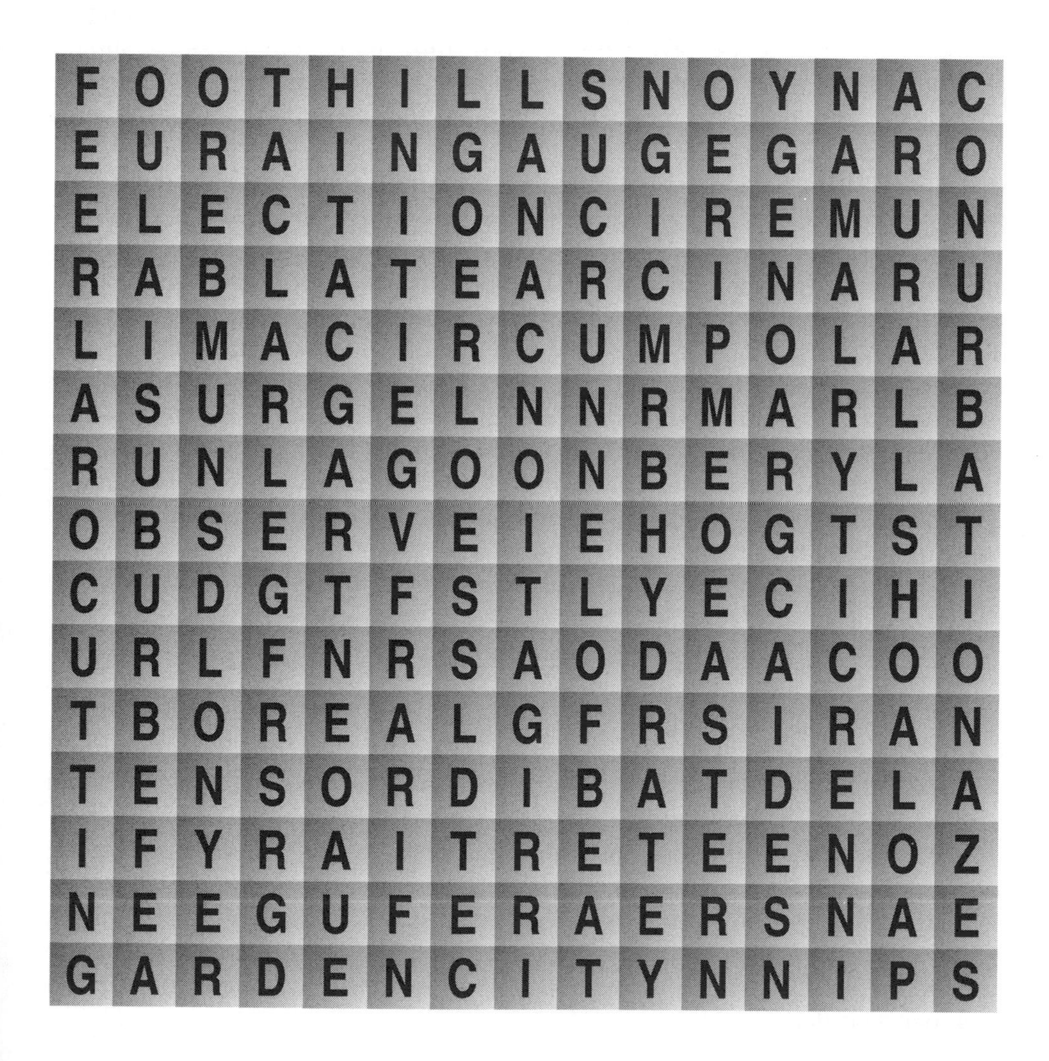

3 MAXIMISING FREE TIME

It isn't what you know that matters, it's what you think of in time.

University life is different from school life and employment. There is lots of free time for skysurfing, street luge, acting in *Trainspotting*, being elected Union Secretary, playing the lute and socialising, but many students find meeting coursework deadlines difficult. Time management techniques are especially vital for those with a heavy sport or social programme, a part-time job in term-time, and part-time and mature students with a job running in parallel with the degree. Developing your time management skills should allow you to do all the boring tasks efficiently, like laundry and essays, leaving free time for other activities. It is unlikely that any one idea will change your life overnight, but a few time saving short cuts can relieve pressure. Use reflection and evaluation skills to identify what to do next and to assign time to your studies.

Ideally one can envisage research, whether for an essay or a dissertation, moving linearly from inception to final report or presentation (see Figure 3.1A). Regretfully, the research process is rarely this simple. The normal elements of life intervene, and the way you view and treat any topic usually changes as research progresses. This means that a linear research model is not realistic. The reality of normal progress, (Figure 3.1B) requires plenty of time for the research process to evolve. Half-way through your research you may have to go back almost to the start, reconsider your approach and execute a revised programme. Increasing your ability to manage your time, and recognising and adjusting to changing goal posts, are vital skills improved by university.

3.1 Is there a spare minute?

Start by working out what time is available for research and study by filling out your timetable using **Try This 3.1**. Assume social and sport activities will fill every night, and all weekend and that arriving at University before 10 is impossible. The remaining time is available for research, reading, thinking, planning and writing without touching the weekend or evenings. If you add a couple of evening sessions to the plan it will save money, due to temporary absence from bar or club, and get essays written. Divide this total ~~free~~ research time by the number of modules to get a rough target of the hours available for support work per module.

Figure 3:1 The research process

TRY THIS 3.1 – What spare/research time?

Fill in your timetable: lectures, practicals, ... the works. Block out an hour for lunch and a couple of 30-minute coffee breaks each day. Add up the free hours between 10 and 5 to find your Total Research Time.

	Morning		Afternoon		Evening
Monday					
Tuesday					
Wednesday					
Thursday					
Friday					
Saturday					
Sunday					

3.2 What do I do now?

Confused? You will be.

Diaries and timetables

University timetables can be complex, with classes in different places from week to week. So diaries are vital. A weekly skeleton timetable will locate blocks of time for study (**Try This 3.1**). Use it to allocate longer free sessions for tasks that take more concentration, like writing, reading and preparing for a tutorial or workshop. Use shorter, 1 hour, sessions to do quick jobs like tidying files, sorting lecture notes, summarising the main points from a lecture, reading a paper photocopied for later, highlighting urgent reading, on-line searches, thinking through an issue and making a list of points that you need to be clearer about. Don't be tempted to timetable every hour. Leave time for catching up when plans have slipped.

Lists

Sort out what you need to do under four headings: Urgent now, Urgent next week, Weekly and Fun (see Table 3.1). If you tackle part of the non-urgent task list each week, you will be less overwhelmed by Urgent now tasks at a later date. Have a go at **Try This 3.2**.

Urgent now	Urgent next week	Weekly	Fun
Essay: *Urban Poverty*	Read for tutorial: *Nutrient cycling*	Cliff-diving	Friday: bungee jumping
By Friday	Find out about *Leeds Development Corporation*	Ironing	Party: Dan's birthday get card and bottles
Report 3 for GIS prac.		Supper	

Table 3:1 Keeping track of essentials

TRY THIS 3.2 – Essential or not?

Do a quick version of Table 3.1 for the next three weeks. Put a * against the items that you want to do in the next four days, and make a plan.

Photocopying a diary template with your regular commitments marked: lectures, tutorials, sport sessions, club and society meetings, gives a weekly skeleton for planning. If weekly planning is too tedious, go for the 30-second breakfast-time, back-of-an-envelope version. It can really assist on chaotic days when classes for one hour spread across the day encourage time to disappear. There are free hours but 'no time to do anything properly'. Completing short jobs will avoid breaking-up days when there is more time. Try to set your day out something like this.

9 Lecture	10 *Coffee Anne and Dan*	11 Lab. Computer Practical	12 *Finish computer practical Lunch*	2 Tutorial	3 *Sort file. Read last week's Africa seminar notes.*	4 Africa Seminar	5–9 *Shop. Night Out*

Even on average days which are easier to manage two or more free hours give more research time for concentrated activities. Set out your average day like this.

10–12 *Notes for Tutorial essay Victorian housing*	12 Lecture	1 *Lunch e-mail*	2 Computer Practical	3–5 *Notes Tutorial essay Victorian housing*	5–9 *Telly and 'phone calls.*

Knowing what you want to do in your research time saves time. Deciding in advance to go to the library after a lecture should ensure you head off to the right floor with the notes and reading lists you need. Otherwise you emerge from a lecture, take 10 minutes to decide you would rather read about urban redevelopment than sustainability, discover you haven't got the urban reading list, so look at the sustainability list to decide which library and floor to visit. All this takes 45 minutes and the time has gone.

3.3 Tracking deadlines

Deadlines are easily forgotten. For some people a term or semester chart will highlight deadlines that initially seem far away. Figure 3.3 shows two chart styles. Which would suit you? The first is essentially a list, whereas the second one shows where pressure points build up. In this example Weeks 10 and 11 already look full. The computer report due on Friday needs finishing before the Ball on Thursday! This second style highlights weeks where personal research time is limited by other commitments.

Module	Assessment	Due Date	Personal Deadlines
GEOG1030 Landscape Processes	Essay: Slope Instability Factors	10 Dec.	*Check out a couple of background texts and case studies by November 25. Draft by Dec. 1, diagrams and revise for Dec. 10*
GEOG1070 Statistics	Practical	16 Nov.	*Sort out the data set and run EXCEL program by 12 Nov. Write up by 15 Nov*

Week	Social	Workshops	Computing	Tutorial	Essays	Laboratory
1 Sept. 29						
2 Oct. 6			Report Fri.			
3 Oct. 13		GEOG1010 Tues. Worksheet				Climate Report Thurs.
10 Dec. 1	Geog Soc. Ball Thurs.	GEOG1060 Worksheet Mon.	Report 6 Fri	Tues. Oral Presentation: Aquaculture		
11 Dec. 8	End of Term Xmas Shop Hall Dinner	GEOG1060 Summary Test Tues.	Report 7 Fri.		National Parks essay: Fri.	Sediment practical Thurs.

Figure 3:2 Sample semester planners

Essay planning

Run it past yourself, backwards. Assuming it is due in seven weeks time. Allow:

(a) Week 7: slippage, the worst flu ever, final checking and completing references and diagrams (This is generous, unless you really get flu.)

(b) Week 6: Finish final version.

(c) Weeks 4 and 5: Read and draft sections, review and revise, repeat library and electronic searches.

(d) Week 2–3: Brainstorm keywords, do library and Internet search, decide on main focus, highlight lecture notes and references, browse for additional information, start writing.

(e) Week 1: Put the main jobs and deadline dates on your semester plan and Urgent list (Table 3.1).

OK, the chances of 1:1000 students doing this are small, but it is a good idea!
See Chapter 18 on dissertations. Starting to plan dissertations in year 1 is a good idea!, though not essential.

3.4 What next?

Get into the habit of reviewing what you have to do and look at the relative importance of different activities, so you don't miss a deadline. Have a go at **Try This 3.3** as practise in prioritising.

Then reflect on where you could re-jig things to release two lots of 20 minutes. Twenty minutes may not seem much, but if you grab a couple of slots to sort notes, reread last week's lecture notes or skim an article, that's 40 minutes more work than you would have done.

TRY THIS 3.3 – Priorities?

Using yesterday as the example, jot down the time devoted to each task yesterday, amending the list to suit your activities.

Real Life	Hours	Priority	Geography Degree	Hours	Priority
Eating			Reading		
Sleeping			Browsing in the library		
Washing/Dressing			Lecture attendance		
Exercise			Sorting lecture notes		
Travelling			Writing		
TV			Thinking		
Other leisure ...			Computer Practicals		
Reading for fun			Laboratory Practicals		
Housework			Planning Time		
Washing/Ironing			On-line searches		

Tackling **Try This 3.3** might encourage you to use a day planner to organise your time better, similar to the one shown in Table 3.2. To have a go write a 'to do' list for tomorrow, then order your activities. (2 = indicates an intention to be double tasking, in this case reading while the washing tumbles around.) Put some times against the activities. Ticking off jobs as they are done feels good!

If you can do a task immediately and easily, that may be the most efficient approach. Generally it helps to allocate larger tasks to longer time chunks and

Priority	MUST DO!	When
1	Post Mother's Day card	On way to college
2 =	Read Chaps 3–5 of *Wave Power* by Sue Nami	10–12
3	Look at ecology notes for seminar	12–1
2 =	Launderette visit	3–5
4 =	Check e-mail	After lunch
4 =	Spell check tutorial essay	After supper
6	Lecture 12.00 in Geography Lecture Theatre	After lunch
5	Sort out computer practical notes	After supper
7	Make a list of jobs for tomorrow	After supper

Table 3.2 An organiser like this?

leave little tasks for days that are broken up. Do not procrastinate: 'I cannot write this essay until I have read ...' is a lousy excuse. No one can ever read all the geographical literature, so set a reading limit, then write, and go clubbing.

Top Tips

☺ Get an alarm clock / buzzer watch / timetable / diary.

☺ If the geography library is full of your best mates wanting to chat, head to the Music or Biophysics library. Do tasks requiring total concentration in comfortable conditions where the lighting is good, the atmosphere conducive to study and no-one will interrupt you.

☺ Plan weekends well ahead; sport, socialising, and shopping are critical. Having worked hard all week, you need and deserve time off. Following a distracting, socially rich week, maybe there is time for some study. Sunday can be a good time to draft a report, and a great time for reading and thinking: very few people interrupt.

☺ Filing systems: 'so many modules, so many handouts, my room could be a recycling depot'. Take 10 minutes each week to sort out notes and papers. Supermarkets let you have strong boxes free. Indexing files will help.

☺ Investigate the use of bibliographic databases to organise your references. Are you exploiting your IT skills to save you time? Computers and especially the Internet encourage a positive feeling of hours spent diligently communicating with the universe. It also takes hours. Are you being side-tracked? Ask yourself 'Am I wasting good living time?'

☺ Good (OK, some or any) organisation can save on stress later. Being stressed usually wastes time. Not all assignments are easy. Recognise that the difficult ones, and especially those everyone dislikes, will take longer and therefore plan more time for them. Divide the tasks into manageable chunks and tackle them separately. Finishing parts of a task ahead of time gives you more opportunity to think about the geographical interpretations.

☺ Vacations. Recover from term. Have a really good holiday. Consider taking a typing course, look at a speed-reading guide, or follow an on-line tutorial to improve IT and GIS skills? Think about dissertation possibilities.

☺ Time has a habit of drifting away very pleasantly. Can you spot and limit lost time when the pressure is on? Minimise walking across campus. Photocopy at lunchtime when you are in the Union anyway. Pick up mail while in the department. Ask yourself 'Is this a trip I need to make?', 'Could I be more time efficient?' Make an agreement with a friend to do something in a certain time and reward yourselves for success afterwards.

☺ View apparently 'dead time', when walking to university or cleaning the bathroom, as a 'thinking opportunity'. Use it to plan an essay, mentally review lecture ideas ...

☺ Be realistic. Most days do not map out as planned, things (people) happen, but a plan can make you a little more efficient some of the time.

If you decide to investigate some of these ideas, give them a real go for three weeks. Then re-read this chapter, consider what helped and what did not, and try something else. Find a routine that suits you and recognise that a routine adopted in first year will evolve in following years. A realistic study timetable has a balance of social and fitness activities. Don't be too ambitious. If there was no reading time last week, finding 30 minutes to read one article this week is a step forward.

3.5 References and further reading

Kneale, P.E. 1997 Maximising Play Time: time management for geography students, *Journal of Geography in Higher Education*, Directions, **21**, 2, 293–301.

Rudd, S. 1989 *Time Manage Your Reading*, Gower, Aldershot.

Island links

By changing one letter at a time and keeping to real words, can you move around the islands? Answers on p.253.

B	A	L	I		M	U	L	L		L	O	N	G		B	U	T	E
M	U	C	K		E	I	R	E		U	I	S	T		F	A	R	N

Are there other views on this?

4 LIBRARY AND ELECTRONIC RESOURCES?

And therefore education at the University mostly worked by the age-old method of putting a lot of young people in the vicinity of a lot of books and hoping that something would pass from one to the other, while the actual young people put themselves in the vicinity of inns and taverns for exactly the same reason.

(Pratchett 1994, *Interesting Times*)

Once you discover all the lecture notes that you took so conscientiously are completely unintelligible, or you did not quite make it to a lecture, using the library might be a good wheeze. Inconveniently, many libraries have texts that geographers need catalogued under agriculture, sociology, law, politics and civil engineering. This usually means geography texts are found at many locations and sometimes in different buildings. Some geography departments have their own collections, and those in other departments, in geology, philosophy or chemistry for example, may be useful. Then there are all the electronic resources. Material can be accessed world-wide. Such fun. There is a maze of information, but finding the way around is not always obvious. While it may be beyond Mulder and Scully, you CAN DO THIS. Skills employed include researching, evaluation, information retrieval, IT, flexible thinking and scheduling.

University libraries can seem scary and confusing. Most people feel very lost for the first few visits. This chapter gives information about library resources and research strategies, tips and hints that will, hopefully, reduce the mystery.

4.1 Library resources

For most library visits you need a library card to get in and out, cash or card for photocopying, paper for notes, and watch your bags; the opportunist thief finds a library attractive, people leave bags while searching the shelves. A few minutes with guided tours, watching videos and on-line explanations of your library's resources, and tips on accessing library and on-line documents will save hours of inefficient searching. Use the library staff. Ask them to show you how the catalogues and search engines work. It is probably harder for geographers, than historians or geologists, to find their way to the right material. Find out where newspapers, the collections for economics, sociology, languages, politics, education . . . , and the collections for your option subjects are kept.

Books and journals

There is an emphasis, at university, on reading academic journals. Journals contain collections of articles written by experts, published in every area of academic study. They are the way in which academics communicate their thoughts, ideas, theories and results. The considerable advantage of a journal over a book is that its publication time is usually six months to two years. Recent journals contain the most recent research results. Check the location of:

- ✔ Recent issues of journals or periodicals. These may be stored in a different area of the library. At the end of the year they will be bound and join the rest of the collection. Reading recent issues can give a real feel for the subject and topics of current, research interest.
- ✔ Government publications with endless tables of vital information for a geographer.
- ✔ Oversized books, they do not readily fit on shelves, and are often filed as Quartos – at the end of a subject section. They are easy to miss.
- ✔ Stack collections containing less commonly used books and journals.

Catalogues

Cataloguing systems are exclusive to particular universities. Happily, every library has handouts about how to retrieve material. Library information is accessed either via an on-line computer catalogue, which shows where the book should be shelved and whether the copies are on loan or not, or from a card index. Before searching, highlight the papers and books on the reading list that you want to read so search time is quick. If the books are out, check at the shelf references for other texts that will substitute. If the on-line catalogue is accessible from any networked campus computer, you can do bibliographic searches, and mark up reading lists while the library is shut. Traditional card index catalogues locate texts by author and title, but a keyword search is not possible.

How do you know which items on a reading list are journals and which are books?

There is a convention in citing references, used in most texts and articles, that distinguishes journal articles from books, and from chapters in edited books. Traditionally, a book has its TITLE in italics (or underlined in hand-written text), a journal article has the title of the JOURNAL in italics, and where the article is a chapter in an edited book the BOOK TITLE is in italics:

Ahnert, F. 1998 *Introduction to Geomorphology*, Arnold, London.

Purvis, M., Drake, F., Clarke, D., Phillips, D. and Kashti, A. 1997 Fragmenting uncertainties, some British business responses to stratospheric ozone depletion, *Global Environmental Change*, **7**, 2, 93–111.

Cosgrove, D. 1997 Prospect, perspective and the evolution of the landscape idea, in Barnes, T. and Gregory, D. (eds.) *Reading Human Geography, the poetics and politics of inquiry*, Arnold, London, 324–342.

For example, to find the Cosgrove article, search for Barnes and Gregory (1997). This convention eases library searches because, as a rule of thumb, you search for the italicised item first. You will never find the title of a journal article in a library main catalogue, but you will find the journal title and its library shelf location. In spotting journals look for numbers, as here **7**, 2, 93–111, indicating volume 7, issue 2, and pages 93–111; books do not have this clue. Where there are no italics the game is more fun, you have to work out whether it is a journal or book you are chasing. (All students play this game; it's a university tradition.)

Unfortunately you cannot take out all the books at the beginning of term and keep them for the whole term. Find out what you can borrow and for how long, and what is available at other local libraries, the city or town library? If a library does not hold the article or book you want, you can borrow it from the British Library via the inter-library loan service. There is likely to be a charge for this service, so be sure it is a book or article you really need. Many university libraries offer short loan arrangements for material that staff have indicated everyone will want to read. Check out the system, especially the time restrictions on a loan. Return your books on time. *Fines are serious*, especially for restricted loans, and a real waste of good drinking money. When you need to get your hands on texts that are out, RECALL them. It encourages other people, especially staff, to return them.

On-line searches can be made using the title, author or keywords. Before searching make a list of keywords, and decide if you need to search for English and American spellings. Searching for 'mountain bikes erosion' yielded 716,975 hits, far too many. Boolean operators will speed up and refine the search, by cutting out irrelevant sites (see Figure 4.1). Entering 'Mountain+bike+erosion' yielded 1134 hits, which is still too large. If you refine your area of interest again, as in 'Mountain+bike+erosion+National+Park+United+Kingdom' you hit 21 sites. A manageable research list.

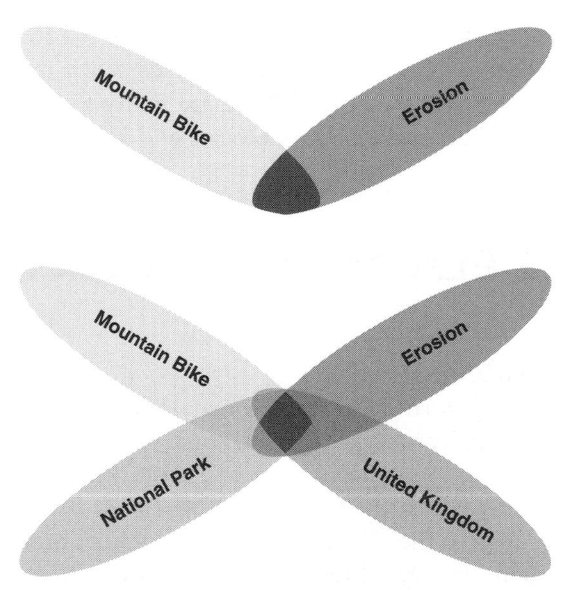

Figure 4:1 Refining searches with Boolean operators

There are three main Boolean operators used either in words or symbol form: + , – or **AND OR NOT**. 'bike, bicycle' should pick up entries for both. 'Migration **OR** transmigration **AND** gender **NOT** Latin America' should locate material on migration and gender issues from areas other than Latin America. Use **OR** when there are synonyms, and **NOT** to exclude topics. The introduction or Help information for each electronic database should explain which symbols can be used for searches.

Using root words, like Environ*, will find all the words that have environ as the first 7 letters, like environment, environmental and environs. Beware using root* too liberally. Poli* will get politics, policy, politician, political which you might want, and policeman, polite, polish and Polish, which you might not.

American and English spellings can be a nightmare, use both in keyword searches. Here is a starter list but add to these as you find them. Beware use of -ise (UK) and -ize (US) endings, too.

English	American	English	American	English	American
artefact	artifact	defence	defense	metre	meter
behaviour	behavior	dialogue	dialog	mould	mold
catalogue	catalog	draught	draft	plough	plow
centre	center	enclose	inclose	sulphur	sulfur
cheque	check	enquire	inquire	traveller	traveler
colour	color	foetus	fetus	tyre	tire
counsellor	counselor	labour	labor	woollen	woolen

There is also the geographical problem of changing place-names. Some are in common use: Peking and Beijing, Ceylon and Sri Lanka, Burma and Myanmar, Canton and Guangzhou. City names, especially those in the former Soviet Union, may require careful cross-checking through atlases, with map curators or *The Statesman's Year-Book* (Hunter 1997), and there are many European examples that can cause confusion, Lisbon and Lisboa, Cologne and Köln, Florence and Firenze, for example.

4.2 Electronic resources

Bibliographic databases

These databases contain information about journal publications and some include book details. There is usually author, title and source details and, in most cases, an abstract or short summary. Databases may be networked, held in a library CD-ROM collection, or accessed via the Internet. At best, only a few CD-ROMs will be available to you, because of the expense. Don't give your librarians a hard time if you cannot access them. In 1998, Meteorological and Geoastrophysical Abstracts costs $5100 with updates costing $1400, so CD-ROMS are not necessarily a good buy for a library. Searching databases can locate paper titles and abstracts, but are not a substitute for reading the whole paper.

For UK students, **BIDS** (Bath Information and Data Service) is a goldmine and user-friendly. It holds details of academic papers in many journals. Check whether your library is networked to BIDS, obtain a login name, password, and get on-line. You can search by authors and keywords. Explore the different options and menus to become familiar with what BIDS has to offer. Mark items of interest as you search, to download via e-mail to your own filespace.

Here is a selected list of databases that are of interest to geographers. What sources can your library access?

AQUALINE covers the hydrological cycle and includes wastewater and sewage treatment references.

GEOREF is the geology resource.

ESPM (Environmental Sciences and Pollution Management) includes aquatic pollution, bacteriology, ecology, energy resources, environmental biotechnology, environmental engineering, environmental impact statements, hazardous waste, industrial hygiene, microbiology and risk assessment.

METEOROLOGICAL AND GEOASTROPHYSICAL ABSTRACTS offer citations for world-wide meteorological and geoastrophysical research, 1970+.

POPLINE has information on demography, family planning services, maternal and child health and related health issues.

PSYCLIT holds information from Psychology journals.

SOCIOFILE has citations and abstracts from journals in Sociology.

TRANSPORT is a bibliographic database for transport research and economic information.

TROPAG and RURAL is a database for agriculture in tropical and subtropical regions including rural development, economic policy and planning

WASTEINFO holds bibliographic information on recycling, waste treatment, waste disposal, waste minimisation, environmental legislation and policy, and associated environmental issues.

World-Wide Web (www)

The Web promises to be the most time-wasting, but fun, element of a degree, while giving the comforting feeling of being busy on the computer all day. Use a keyword list (inner city deprivation, pollution, protected species) and Boolean logic (inner+city+deprivation) to exclude unwanted sites. Moving the cursor over highlighted text (usually blue) and double clicking the mouse will link to other documents. This is known as a *hypertext link*. Book marking 'favourite' pages will save you having to search from scratch for pages you use regularly. You should be able to e-mail documents to your own filespace, or save to floppy disc. On some computers you can open www and word processing packages simultaneously, and cut and paste between the two (but be aware of plagiarism issues, see section 12.6).

Major problems can arise if you print files with pictures or graphics. Make sure the printer will handle graphics, or you will clog printer queues and run up an enormous printing bill. The www response rate will be slow at some times of day. Be patient. Look around campus for the faster, newer machines. Check out some sites using **Try This 4.1**.

TRY THIS 4.1 – WWW resources for geographers

Explore some of these sites. If a site address is defunct, use a search engine and the site title to locate the updated address, e.g. Association+Geographic+Information.

The Association for Geographic Information at http://www.geo.ed.ac.uk/root/agi/agi.html (Accessed 10 January 1999)

British Geological Survey at http://www.bgs.ac.uk/ (Accessed 10 January 1999)

BBC news on-line at http://news.bbc.co.uk/hi/english/uk/ (Accessed 10 January 1999)

US Census Bureau at http://www.census.gov/ (Accessed 10 January 1999)

The ID21 (Information for Development in the 21st century) has digests of social and economic research in Developing Studies, at http://www.id21.org (Accessed 10 January 1999)

The Africa library at http://www.africalibrary.org (Accessed 10 January 1999)

Latin America information at http://www.lanic.utexas.edu (Accessed 10 January 1999)

WWW Virtual Library for Geography at http://www.icomos.org/WWW_VL_Geography.html (Accessed 10 January 1999)

Environmental information for geotechnical, environmental, hydrogeology, geology, mining and petroleum topics at http://www.geoindex.com (Accessed 10 January 1999)

The UK Meteorological Office at http://www.meto.govt.uk/home.html (Accessed 10 January 1999)

Internet Resources on the Middle East at http://www.fas.harvard.edu/~mideast/inMEres/inMEres.html (Accessed 10 January 1999)

New Scientist at http://www.newscientist.com (Accessed 10 January 1999)

The (London) *Times* at http://www.the-times.co.uk (Accessed 10 January 1999)

The United Nations Department of Economic and Social Affairs, Statistics Division at http://www.un.org/Depts/unsd/ (Accessed 10 January 1999)

Desert Research Institute at the Quaternary Sciences Centre http://www.dri.edu/QSC/ (Accessed 10 January 1999)

Water Quality Sites on the Internet at http://seaborg.nmu.edu/water/Default.html (Accessed 10 January 1999)

Geomorphology from Space (1986) a text by N.M. Short and R.W. Blair is now out of print but can be found at: http://daac.gsfc.nasa.gov/DAAC_DOCS/geomorphology/GEO_HOME_PAGE.html This site shows satellite images with some terrestrial material, maps, and photos. (Accessed 10 January 1999)

London Research Centre Home Page, information about urban affairs at http://www.london_research.gov.uk/index.html

⚠**Warning: Try This 4.2 could be totally useful or utterly frustrating.** It is included because it will be useful to some people, but it may be obsolete by April of 1999, everything changes fast in this area. Most were accessed via http://dialog-carl.thames.rlg.org, with an authorised login and password. Ask your librarian what is available to you. Do not get frustrated, libraries and departments cannot possibly afford to pay for access to all these sites. Not being able to access a specific item will not cause you to fail your degree.

⚠**Warning:** The fact that a database exists is no guarantee that it holds the information you need.

TRY THIS 4.2 – www addresses

To develop familiarity with the www and library resources, see how many of these databases you can access. Update the list with current http addresses. Have a brief look at the page that describes the contents of the database, is it one worth bookmarking for future use? Can you add databases to this list? The addresses are given on p.253, but attempt to find them yourself.

If you search for AGRICOLA you will hit hundreds of vineyard and agriculture sites in Italy. Use AGRICOLA+database, and it should be the first hit.

AGRICOLA	Worldwide information on agriculture and related fields, 1970+.
Asian studies	Covers business, economics, and new industries of Pacific Rim nations, 1985+.
CNN Interactive	The web site for CNN (Cable News Network)
CRIS/USDA	Research in agriculture, food and nutrition, forestry, and related fields.
FedWorld Information Network	US government and business information.
GeoArchive	Covers geoscience, hydroscience, and environmental science, 1974+.
GeoRef	Covers world-wide technical literature on geology and geophysics, 1785+.
Handbook of Latin American Studies	Covers Latin American Studies ,1990+.
ITAR/TASS	News from the official state news agency for Russia, 1996+.
Reuters	World-wide newswire, information for the last year
Russian Academy of Sciences Bibliographies Indexes	Slavic materials, 1992+.
South China Morning Post	English-language news from Hong Kong, 1992+
Wilson Social Science Abstracts	Abstracting and indexing of 415+ English language social science journals, 1988+.

⚠ **Warning:** www documents may be of limited quality, full of sloppy thinking and short of valid evidence. Some are fine, but be critical. Anyone can set up a www site. Look for reputable sites especially if you intend to quote statistics and rely heavily on site information. Government and academic sites should be OK.

⚠ **Warning:** Think about data decay, which time period does the information relate to? If, for example, you have economic or social data from the 1980s for Bosnia, or the old USSR it will be fine for a study of that period in those regions, but of negligible value for a current status report. Check dates.

⚠ **Warning:** Do not plagiarise. You can cut and paste from the Internet to notes, but if you cut and paste to an essay the source must be properly acknowledged. See sections 12.6 and 14.3.

E-journals

Some geographical journals are available on-line and the numbers will expand. Conveniently, the flagship British geography journal, *Transactions of the Institute of British Geographers*, 1935–1994, is at: http://www.rgs.org/pu/8publetr.html (Accessed 15 January 1999). The issues in the most recent five-year period can be found via BIDS. There is a list of geographical journals with electronic connections at http://ppt.geog.qmw.ac.uk/tibg/tibg_lnk.html. Subscription only journals will not be accessible, but some have abstracts that can be viewed at no expense.

Check out the other geographical journals available electronically through your library, and remember that journals from other subject areas may be relevant, such as Sociological Research On-line at
http://www.socresonline.org.uk/socresonline/ (Accessed 9 January 1999).

4.3 Research strategies

There are oodles of background, research documents for just about every geographical topic, usually far too many. The trick in the library is to be efficient in sorting and evaluating what is available, relevant, timely and interesting.
A library search strategy is outlined in Figure 4.2. Look at it carefully, especially the recommendations about balancing time between searching and reading. Library work is iterative. Remember that on-line searches can be done when the library is closed but computer laboratories are open. Become familiar with your local system, use **Try This 4.3** and **Try This 4.4** as a starting point. Good library research skills include:

- ✔ Using exploration and retrieval tools efficiently.
- ✔ Reading and making notes.
- ✔ Evaluating the literature as you progress.
- ✔ Recording references and search citations systematically, so referencing or continuing the search at another time, is straightforward.

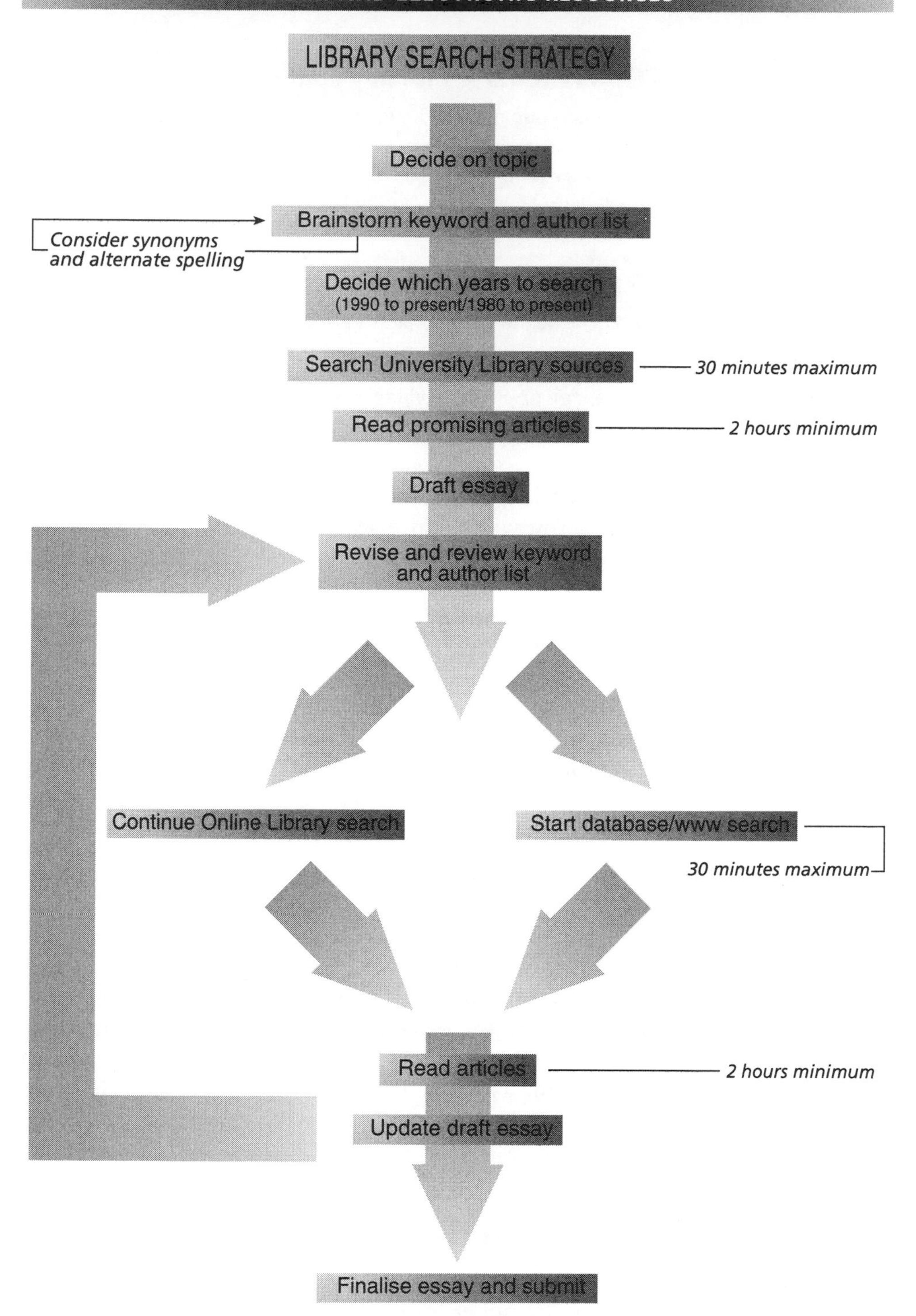

Figure 4:2 Library search strategy

TRY THIS 4.3 – Library search

Choose any topic from one of your modules. Make a list of three authors and six keywords. Do a search to explore the papers and texts available in your library. Compare the results with the reading list. Is there a paper that should be followed up which is not on the reading list?

TRY THIS 4.4 – Journal search

Use BIDS and databases to check the authors and keywords used in **Try This 4:3**, but only note the journals in your library, those you can read later.

Set a maximum of 30 minutes for on-line searching, then read for at least two hours.

Beware the enticements of on-line searches. It is possible to spend all day searching electronic sources. You will acquire searching skills, know there is a paper with the ideal title in a library in Australia or a foreign language and have nothing for an essay.

Ignore enticing www gateways during initial research. For most undergraduate essays and projects the resources in the library are adequate; search and read these articles first. Look at wider resources later, after the first draft is written, and only if there is bags of time. Web searches are more appropriate for dissertation and extended project work. You cannot access and read everything, essays have short deadlines, and reading time is limited. The trick is to find documents available locally and at no cost.

Reading takes time

Reading lists

Some lecturers give long reading lists with lots of alternative reading. This is vital where members of large classes want to access documents simultaneously. Other lecturers give quite short reading lists, especially at levels 2 and 3. They may include essential reading items, and a list of authors and keywords. This approach allows each student to explore the available literature independently. Where the reading lists are of the first type, it is wise to view it as a big version of the second! Use **Try This 4.5** to explore an author's work.

TRY THIS 4.5 – Author search

Take one geography author whose name appears on a reading list and design a library search to explore what that author has written. (Hint: include geo-databases, BIDS and the social science or science citation indexes.)

Review articles

Review Articles, especially those in *Progress in Physical Geography* and *Progress in Human Geography* can give an illuminating synthesis of the recent literature, and point you to other references. Some journals have themed issues. For example, in 1997 *Environment and Planning A* had four themed issues on Citizenship and International Migration, Multilevel Modelling, Cultural Industries and India 1947–1997. You may start out with one reference but in a themed volume there will be papers on inter-related topics that are, at the least, worth browsing.

> Reserve popular books – and read them!

Linked reading

If you happen to read Cosgrove (1996) you will quickly realise you need to read Thrift (1996). It may seem obvious, but having read article A you may need to read article B to understand A. Cross-referencing is a normal process. Take accurate notes of the references to follow up.

Be particularly critical of sources which may have a bias or spin. As Vujakovic (1998) points out, mass media materials are particularly vulnerable to personal or biased reporting. Use **Try This 4.6** to keep track of your reading, and a balance between electronic searching, (30 minutes) to real reading (2 hours.)

TRY THIS 4.6 – Self-assessment of a library search

Use this grid to keep a tally of sources and authorities when doing a library search or preparing a report or essay. Check for an advantageous balance of recent citations and that all the appropriate sources are used, in addition to those on the reading list.

Sources Used in Search for...

Books	Journal Articles	Government Reports	www Sites	Other

Number of papers NOT ON and ON the reading list:

None	1	2–4	5–7	7–10	11+
/	/	/	/	/	

Number of papers from each time period:

Pre-1980s	1980–1984	1985–1989	1990–1994	1995–1999	2000–date

4.4 Why am I searching?

Library searches are never done in isolation. Before starting, review your reasons for searching, and focus your keyword search. Put a limit on your searching time and on the types of document to include. Suggestions on this front include:

- ✔ Module essays; start with the reading list, and only explore further when you have an initial draft. Look critically at the gaps in your support material and use **Try This 4.6.**
- ✔ 'State of the Art Studies' exploring the current state of knowledge on a topic; limit searches to the last 2–5 years.
- ✔ An historical investigation of the development of an idea considering how knowledge has changed over 10, 20, or more years; aim for a balance between the older and newer references.
- ✔ A literature review should give the reader an outline of the 'state' of the topic. It may have a brief historical element, mapping the development of the subject knowledge, leading into a more detailed resume of research from the past 5–10 years.

Do not forget international dimensions, the topic might be theoretical (urban development, social housing, migration, urban climatology) but there may be regional examples that are worth considering, so check out the journals of other countries. The *Singapore Journal of Tropical Geography*, *The Canadian Geographer*, *Annals of the Association of American Geography* and *Irish Geography* are international journals, but also contain articles reflecting national and regional concerns.

WRITE UP AS YOU GO, keep noting and drafting, and keep a record of references in full. Remember to add the reference, dates and pages on photocopies.

4.5 References and further reading

Cosgrove, D. 1996 Windows on the City, *Urban Studies*, **33**, 8, 1495–98.

Hunter, B. (ed.) 1997 *The Statesman's Year-Book, a statistical, political and economic account of the states of the world for the year 1997–1998*, (134th edn), MacMillan Reference, London.

Livingstone, I. and Shepherd, I. 1997 Using the Internet, *Journal of Geography in Higher Education*, Directions, **21**, 3, 435–43.

Pratchett, T. 1994 *Interesting Times*, Victor Gollancz, London.

Thrift, N. 1996 New Urban Eras and Old Technological Fears: reconfiguring the goodwill of electronic things, *Urban Studies*, **33**, 8, 1463–93.

Vujakovic, P. 1998 Reading Between the Lines: using news media materials for geography, *Journal of Geography in Higher Education,* Directions, **22**, 1, 147–55.

Winslop, I. and McNab, A. 1996 *The Student's Guide to the Internet*, Library Association Publishing, London.

For updates to www sites see
http://www.geog.leeds.ac.uk/staff/p.kneale/skillbook.html

5 EFFECTIVE READING

I read about the hazards of typhoons, I was so frightened, I gave up reading.

You are 'reading' for a geography degree, so it is not surprising that most time at university should (!) be spent with a book. Everyone reads, the knack is to read and learn at the same time. Pressure on library resources often limits the time books are available to you, so it is vital to maximise your learning. This chapter discusses some reading techniques, asks you to reflect on and evaluate what you do now, and consider what you might do in future! As with playing the trombone, practise is required. There is a vast amount of information to grapple with in a geography degree. Reading, thinking and note making are totally inter-linked activities, but this chapter concentrates on the reading element.

5.1 READING LISTS

Inconveniently, most geography lecturers sort reading lists alphabetically by module, but you need them sorted by library and library floor location. Either use highlighter pen to indicate what is in which library, and take advantage of the nearest library shelf when time is limited, or make a list of the articles/chapters/papers you want to read each week. Now code it by *library* and by *floor* rather than by module (see Table 5.1.) This list needs to be twice as long as you can reasonably do in a week, so if a book is missing there are alternatives.

Author	**Journal/Title**	**Class mark**	**Module**	**Libray/Floor**
Gupta, A. and Asher, M.G. 1998	*Environment and the Developing World*	987.65	Tropical environments.	Floor 6 Science Lib.

Figure 5:1 Library list sorter

Top Tip

Carry Reading Lists at all times.

Reading lists are often dauntingly long, but you are not, usually, expected to read everything. Long lists give you choice in research topics, and especially where class sizes are large and books restricted, lots of options. Serendipity cheers the brain. If a book is On Loan, don't give up. There are probably three equally good texts on the same topic with the same library class number. Reading something is more helpful than reading nothing, and RECALL essential texts.

Use examples that are not in the lecture notes

Students tend to request a module text and feel uncomfortable when a tutor says 'there is no set text'. Even when there is one, it is rarely followed in detail. Reading a recommended book is a good idea, but watch out for those points where a lecturer disagrees with the text. Perhaps the author got it wrong, or our understanding of a topic has moved forward, ideas have changed. You do not have to buy all the set texts, buy as a group and share. Watch the noticeboards for second-hand sales.

Reading in support of lecture modules is the obvious thing to do, but 'Do you read for computing, statistics and laboratory classes?' Certainly, the volume of reading expected for practical modules is less than for lecture modules, but zero reading is not right either. Class activities tend to stress the practical, hands-on elements, BUT you should still allocate time for reading, to understand where practical activities fit with the art of geography. If you don't make the connection between practicals and their geographical applications in levels 1 and 2, you are unlikely to utilise techniques to best effect in projects and dissertations.

Top Tip

READ FOR ALL MODULES!

Photocopying is no substitute for reading – but it feels really, really good.

5.2 Reading Techniques

There is a mega temptation to sit down in a comfy chair with a coffee and to start reading a book at page 1. **THIS IS A VERY BAD IDEA**. With many academic texts, by page 4, you will have cleaned the cat litter tray, done a house full of washing, mended a motor bike, fallen asleep or all four and more. This is great for the state of the house but a learning disaster.

Everyone uses a range of reading techniques, speed reading of novels, skip reading headlines etc., the style depends on purpose. As you look through this section reflect on where you use each technique already. For effective study adopt the 'deep study' approach.

Deep study reading

Deep study reading is vital when you want to make connections, understand meanings, consider implications and evaluate arguments. Reading deeply needs a strategic approach and time to cogitate. Rowntree (1988) describes an active reading method known as SQ3R, which promotes deeper, more thoughtful reading. It is summarised in **Try This 5.1**. SQ3R is an acronym for Survey – Question – Read – Recall – Review. Give **Try This 5.1** a go, it may seem long-winded at first, but is worth pursuing because it links thinking with reading in a flexible manner. It stops you rushing into unproductive note making. You can use SQ3R with books and articles, and for summarising notes during revision. You are likely to recall more by using this questioning and 'mental discussion' approach to reading. Having thought about SQ3R with books use **Try This 5.2**.

Browsing

Browsing is an important research activity, used to search for information which is related and tangential to widen your knowledge. In essence it involves giving a broader context or view of the subject which in turn provides you with a stronger base to add to with directed or specific reading. Browsing might involve checking out popular social science, history, science, and introductory texts. Good sources of general and topical geographical information include *The Economist, New Scientist, New Internationalist*, and the country and investment focus supplements in *The Guardian* and *The Financial Times*. Browsing enables you to build up a sense of how geography as a whole, or particular parts of the subject, fit together. Becoming immersed in the language and experience of the topic encourages you to think geographically.

Scanning

Scan when you want a specific item of information. Scan the contents page or index letting your eyes rove around to spot key words and phrases. Chase up the references and then, carefully, read the points that are relevant for you.

TRY THIS 5.1 – SQ3R

SQ3R is a template for reading and thinking. Try it on the next book you pick up.

Survey: Look at the whole text before you get into parts in detail. Start with the cover, is this a respected author? When was it written? Is it dated?

Use the Contents and Chapter headings and subheadings to get an idea of the whole book and to locate the sections that are of interest to you. First and last paragraphs should highlight arguments and key points.

Question: You will recall more if you know why you are reading, so ask yourself some questions. Review your present knowledge, and then ask what else you want/need to know. Questions like: What is new in this reading? What can I add from this book? Where does this fit in this course, other modules? Is this a supporting/refuting/contradictory piece of information?

Having previewed the book and developed you reasons for reading you can also decide whether deep reading and note making is required, or whether scanning and some additions to previous notes, will suffice.

Read: This is the stage to start reading, but not necessarily from page 1, read the sections that are relevant for you and your present assignment. Read attentively but also critically. The first time you read you cannot get hold of all points and ideas.

On first reading: locate the main ideas. Get the general structure and subject content in your head. *Do not make notes during this first reading*, the detail gets in the way.

On second reading: chase up the detailed bits that you need for essays. Highlight or make notes of all essential points.

Recall: Do you understand what you have read? Give yourself a break, and then have a think about what you remember, and what you understand. This process makes you an active, learning reader. Ask yourself questions like: Can I explain this idea in my own words? Can I recall the key points without re-reading the original text?

Review: Now go back to the text and check the accuracy of your recall! Reviewing should tell you how much you have really absorbed. Review your steps and check main points: Are the headings and summaries first noted the right ones, do they need revising? Do new questions about the material arise now that you have gone through in detail? Have you missed anything important? Do you need more detail or examples? Fill in gaps and correct errors in your notes. Ask where your views fit with those of the authors. Do you agree/disagree?

The last question is 'Am I happy to give this book back to the library?'

TRY THIS 5.2 – SQ3R for papers

How do you adapt SQ3R to read a journal article? Work out a five point plan and try it on the next article you read.

Skimming

Skim read to get a quick impression or general overview of a book or article. Look for 'signposts': chapter headings, sub-headings, lists, figures, read first and last paragraphs/first and last sentences of a paragraph. Make a note of key words, phrases and points to summarise the main themes; but this is still not the same as detailed, deep reading.

What is the hypothesis here?

Photoreading

One of many scanning techniques, Scheele (1993) describes a 'photoreading' method that again requires you to identify your aims before scan reading, and mentally and physically, filing the contents.

When reading ask yourself:

- Is this making me think?
- Am I getting a better grasp of the subject material?

If the answer is no, then maybe you need to read something else or employ a different technique. Reading is about being selective, and it is an iterative activity. Cross-checking between articles, notes and more articles, looking back to be sure you understand the point and chasing up other points of view are all parts of the process. Breaking for coffee is OK and necessary! Talking to friends will assist in putting reading in perspective. Have a go at **Try This 5.3**.

TRY THIS 5.3 – Where do you read?

Where do you do different types of reading? 'I need pen and paper in hand to read and learn effectively?' How true is this for you? Three student responses to the first question are on p. 254. Have a quick look if you find reflecting on this difficult.

Academic journal articles and books are not racy thrillers. There should be a rational, logical argument, but rarely an exciting narrative. Usually, authors state their case and then explain the position, or argument, using careful reasoning. The writer should persuade the reader (you) of the merit of the case in an unemotional and independent manner. Academic writing is rarely overtly friendly or jolly in tone. You may well feel that the writer is completely wrong. You may disagree with the case presented. If so, do not 'bin the book', make a list of your disagreements and build up your case for the opposition. If you agree with the author, list the supporting evidence and case examples.

Get used to spotting cues or signposts to guide you to important points and the structure. Phrases like 'The background indicates ...', 'the results show ...', 'to summarise ...' or see **Try This 5.4** to find further examples.

TRY THIS 5.4 – Spotting reading cues

Look through the book or article you are reading for geography at present, and pick out the cue words and phrases. There are examples on p.254.

5.3 How do you know what to read?

What do I know already?

Reading and note making will be more focused if you first consider what you already know, and use this information to decide where reading can effectively fill the gaps. Use a flow or spider diagram (see Figure 4.2 and Figure 8.1) to sort ideas. Put boxes around information you have already, circle areas which will benefit from more detail, check the reference list for documents to fill the gaps, and add them to the diagram. Then prioritise the circles and references, 1 to *n*, making sure you have an even spread of support material for the different issues. Coding and questioning encourage critical assessments and assist in 'what to do next' decisions.

Be critical of the literature

Before starting, make a list of main ideas or theories. While searching, mark the ideas which are new to you with asterisks, tick those which reinforce lecture material, and highlight ideas to follow up in more detail. Questions to ask include:

- Is this idea up to date?
- Are there more recent ideas?
- How does this paper or idea connect to the main thrust of the essay or argument?
- Do the graphs make sense?
- Are the statistics right and appropriate?
- Did the writer have a particular perspective that led to a bias in writing?
- Why did the authors research this area? Does their methodology influence the results in a manner that might affect the interpretation?

Library, author and journal searches start the process and practise allows you to judge the relative value of different documents. After reading, look at the author and keyword list again. Do you need to change it? Exploring diverse sources will develop your research skills. Reading and quoting sources in addition to those on the reading list may seriously impress an examiner.

Narrow Reading ➜ predictable essays and reports ➜ middling marks

Wide Reading ➜ more creative, less predictable responses ➜ higher marks (usually)

It does not usually matter what you read, or in what order. Read something.

How long to read for?

For most people, two hours is long enough to concentrate on one topic. A short article from *New Scientist* or *The Economist* should take less, but some reading takes longer. With longer documents you need a reading strategy, and take breaks. Use breaks to reconsider the SQ elements of SQ3R and decide whether your reading plan needs amending. If you cannot get involved with a text then it is possibly because you cannot get to grips with the point of the article, or do not know why you should be interested. So STOP READING, and skim the chapter headings, skim your notes, refresh your brain on WHY you are reading and what you want to get out of it.

5.4 Styles of writing

There are differences in writing styles across the geographical literature and some people have difficulty with reading and learning from certain kinds of writing. Styles range from the very direct through to the very discursive. This variety needs to be acknowledged from the start and reading, note making and discussion styles need to be matched to the different styles of writing. Much of the geography literature is technical in tone, characterised by short sentences and an information-rich content. Most of the economic geography literature is in this style, as are most science geography papers. In the more philosophical literature and in cultural geography, for example, you will find writing that is less direct and more discursive. It is the overall tone and style that present the 'message', and analysing every word, individually, for meaning will confuse not help. Look for the broad themes rather than the detail. Writing styles reflect the conversational language and approach adopted in particular sections of the discipline. Becoming familiar with the different languages of geography is part of your geographical training. The following articles exemplify a range of written styles. The trick is to adapt your reading and note-making style to maximise your learning. For scientific, factual style writing see:

Lloyd, P.E. and Dicken, P. 1978 *Location in Space: a theoretical approach to economic geography*, (2nd edn), Harper and Row, London.

For examples of a more philosophical or discursive style see:

Barnett, C. 1998 Impure and worldly geography: the Africanist discourse of the Royal Geographical Society, 1831–73, *Transactions of the Institute of British Geographers* **23**, 239–51.

Clarke, D.B. (ed.) 1997 *The Cinematic City*, Routledge, London.

Redclift, M.R. and Benton, T. (eds) 1994 *Social Theory and the Global Environment*, Routledge, London.

Don't be put off by a writer's style. Make sure you pick up the main message from each section. Be sure you are clear on the supporting and opposing arguments. Are things less black and white, are there parallel arguments? Does the text give one side of the argument? Can you think of another side? Do you need to read something else to balance this author's view?

Top Tips

- If you find certain articles difficult to read, it may be due to unfamiliarity with the topic, its setting and related information, rather than the written style. If an article seems difficult, look at some related, scene-setting materials and then reread the paper.
- Ask yourself: Do I understand this? Ask it at the end of a page, chapter, paper, tutorial, lecture . . . and not just at the end!

5.5 References and further reading

Using on-line searching to locate reading skills texts in your library may produce a long list. Refine the search by excluding TEFL (Teaching English as a Foreign Language) and school level texts.

Buzan, T. 1971 *Speed Reading*, David and Charles, Devon.

Northedge, A., Thomas, J., Lane, A. and Peasgood, A. 1997 *The Sciences Good Study Guide*, The Open University, Milton Keynes.

Reynolds, M.C. 1992 *Reading for Understanding*, Wadsworth Publishing Company, California.

Rowntree, D. 1988 *Learn How to Study: a guide for students of all ages*, (3rd edn), Warner Books, London.

Scheele, P.R. 1993 *The PhotoReading Whole Mind System*, Learning Strategies Corporation, Minnesota.

Field trip conundrum

Work out who did what on the fieldtrip, and what mark they got. Answers on p254.

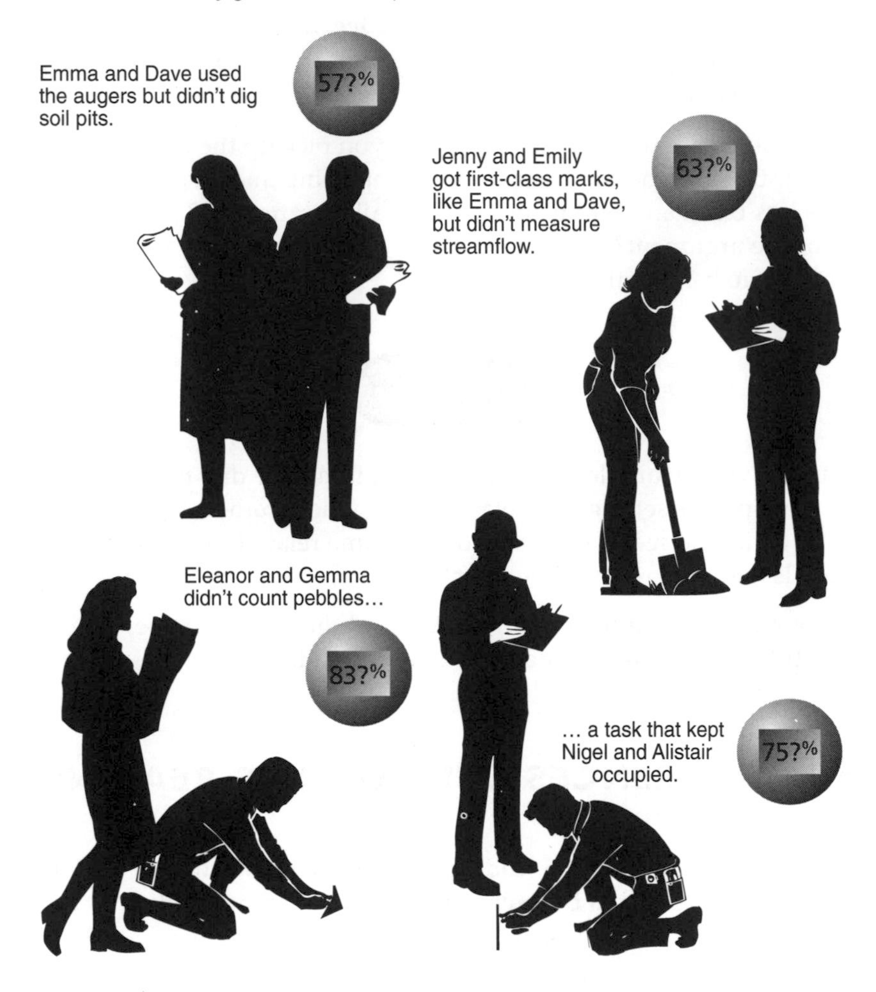

The pebble counters had a 7 in their mark.
Jenny and Emily didn't get 83% and the pebble counters didn't get 75%.

6 MAKING EFFECTIVE NOTES

I made a mental note, but I've forgotten where I put it.

There is a mass of geographical information whizzing around in radio, video and TV reports, specialist documentaries, lectures, tutorials, discussion groups and all that written material including books, journals and newspapers. BUT, just because an article is in an academic journal, in the library, or on a reading list, does not make it a 'Note Worthy' event. Making notes is time-consuming, and ineffective if done on auto-pilot with the brain half-engaged. Note making which lets you learn requires your brain to be fully involved in asking questions and commenting on the ideas. Noting is not just about getting the facts down, it is also about identifying links between different pieces of information, contradictions and examples. Notes should record information in your own words, evaluate different points of view, and encourage the development of your own ideas and opinions. Note taking is a multi-purpose activity, like snow boarding, it gets easier with practice. Good questions to ask when making notes include: 'Is this making me think?' or 'Am I getting a clearer understanding of the topic?'

Many people start reading and making notes without any sort of preview. **A BAD IDEA.** They make pages of notes from the opening section and few, if any, from later in the document. The first pages of a book usually set the scene. Notes may only be needed from conclusion and discussion sections. Sometimes detailed notes are required, but sometimes keywords, definitions and brief summaries are fine. Use **Try This 6.1** to evaluate how the style and length of your note making should change given different types of information, and consider how the SQ3R method (p. 49) fits in your note making process. Then look at **Try This 6.2**, and reflect on what do you do already? What could you do in future?

Reference all quotations

TRY THIS 6.1 – Styles of note making

What styles of notes are needed for these different types of information? There are a couple of answers to kick start ideas (after Kneale 1998).

Academic content	*Style of notes*
Significant article but it repeats the content of the lecture.	*None, it is in the lecture notes, BUT check your notes and diagrams are accurate. Did you note relevant sources, authors?*
Fundamental background theory, partly covered in the lecture.	
An argument in favour of point x.	
An argument that contradicts the main point.	
An example from an odd situation where the general theory breaks down.	
A critically important case study.	
Just another case study.	
Interesting but off the point article.	*A sentence at most! HOWEVER, add a cross-reference in case it might be useful elsewhere.*
An unexpected insight from a different angle.	
An example/argument you agree with.	
An argument you think is unsound.	1. *Brief notes of the alternative line of argument, refs and case example.* 2. *Comment on why it does not work so the argument makes sense to you at revision.*
A superficial consideration of a big topic.	
A very detailed insight into a problem.	

TRY THIS 6.2 – How do you make notes?

Look at this unordered jumble of note making activities and ✔ those likely to assist learning, and put a ✘ against those likely to slow up learning

What do you do already? Are there some ideas here that are worth adopting in the future?

Leave wide margins	Ignore handouts
Identify what is not said	Code references to follow up
Compare and revise notes with friends	Store notes under washing
Do loads of photocopying	Copy big chunks from books
Underline main points	Always note references in full
Use the library for socialising	Make notes from current affairs programmes
Doodle lots	Make short notes of main points and headings
Turn complex ideas into flow charts	Use cards for notes
Ask lecturers about points that make no sense	Order and file notes weekly
Ask questions	Jot down personal ideas
Highlight main points	Share notes with friends
Natter in lectures	Write illegibly
Copy all OHTs	Use coloured pens for different points
Scribble extra questions in margins	Write shopping lists in lectures
Write down everything said in lectures	Annotate handouts
Take notes from TV documentaries	Revise notes within three days of lectures

6.1 MAKING NOTES FROM PRESENTATIONS

Geography lectures and seminars are awash with information so memory meltdown syndrome will loom. Make notes of important points, you cannot possibly hope to note everything. Listen to case studies and identify complementary examples. Highlight references mentioned by the lecturer, and keep a tally of new words. Your primary goal in presentations should be to participate actively, thinking around the subject material, not to record a perfect transcript of the proceedings. Get the gist and essentials down in your own words.

Lecture notes made at speed, in the darkness of a lecture theatre, are often scrappy, illegible and usually have something missing. If you put notes away at once, you may not be able to make sense of them later. Try to summarise and clarify notes within a day of the lecture. This reinforces ideas in your memory, hopefully stimulates further thoughts, and suggests reading priorities.

6.2 Making notes from documents

Noting from documents is easier than from lectures, because there is time to think about the issues, identify links to other material and write legibly the first time. You can read awkward passages again, but risk writing too much. Copying whole passages postpones the hard work of thinking through the material and thereby wastes time and paper. Summarising is a skill that develops given practise. Give **Try This 6.3** a go next time you read a journal article. It won't work for all articles but is a start to structuring note making.

TRY THIS 6.3 – Tackling journal articles

Use this as a guide when reading a journal article or chapter (from Kneale 1998).

1. **First read the article**
2. Write down the reference in full and the library location so you can find it again.
3. Summarise the contents in two sentences.
4. Summarise in one sentence the main conclusion.
5. What are the strong points of the article?
6. Is this an argument/case I can agree with?
7. How does this information fit with my current knowledge?
8. What else do I read to develop my understanding of this topic?

Think about where you will use your information. Scanning can save time if it avoids you making notes on an irrelevant article or one that repeats information you have already. In the latter case, a two-line note may be enough, e.g.

Withyoualltheway (2010) supports Originality's (2005) hypothesis with his results from a comparable study of the ecology of lemmings.

or

Dissenting arguments are presented by Dontlikeit (2010) and Notonyournellie (2010) who made independent, detailed analyses of groundwater data to define the extent of landfill leakage. Dontlikeit's main points are ...

or

Wellcushioned (2010) studying 27 retail furniture outlets in Somerset and Virginia, showed price fixing to be widespread. His results contrast with those of Ididitmyway's (2010) report on prices in Bangkok, because ...

The fuller note may accrue additional marks in an essay or exam. You do not have long to impress an examiner. An essay with one case study as evidence is likely to do less well than one which covers a range of examples or cases albeit more briefly, PROVIDED THEY ARE RELEVANT. Make notes accordingly.

Top Tips

- **New words** Almost every geography module has its specialist vocabulary. Keep a record of new words and check the spelling! The trick is to practice using the 'jargon of the subject' or 'geography-speak'. Get familiar with the use of words like decentralisation, racialisation, hysteresis, and by-passing. If you are happy with geographical terminology, you will use it effectively.

- **How long?** 'How long should my notes be?' is a regular student query, and the answer relates to lengths of string. The length of notes depends on your purpose. Generally, if notes occupy more than 1 side of A4, or 1–5 per cent of the text length, the topic must be of crucial importance. Some tutors will have apoplexy over the last statement, of course there are cases where notes will be longer, but aiming for brevity is a good notion. **Try This 6.1** gave some guidelines.

- **Sources** Keep an accurate record of all research sources and see Chapter 14 for advice on how to cite references.

- **Quotations** A direct quotation can add substance and impact in writing, but must be timely, relevant, fully integrated, and always ensure quotations are fully referenced (Mills, 1994: Chapter 14).

  A quote should be the only time when your notes exactly copy the text. Reproducing maps or diagrams is a form of quotation and again the source must be acknowledged. If you rework secondary or tertiary data in a figure, table, graph or map, acknowledge the source as in 'See figure xxx from Mappit *et al.* (2010)'. 'The data in Table xxx (taken from Plottit 1991) shows that ...', 'Re-plotting the data (Graffit 2010) shows ... (Figure xxx)'.

- **Plagiarism** If you do not fully acknowledge your sources then the university may impose penalties, which can range from loss of marks to dismissal from a course. The regulations and penalties for plagiarism will be somewhere in your University Handbook; something like:

  > Any work submitted as part of any university assessment must be the student's own work. Any material quoted from other authors, must be placed in quotation marks and full reference made to the original authors.

  If you copy directly during note making, you run the risk of memorising and repeating the material in an essay. It is therefore vital to adopt good note making habits to avoid plagiarism. Read and think about ideas and main points, then make notes from your head, using your own language. Find new ways to express ideas. This is not as difficult as it seems, but practice helps. The original author wrote for a specific reason, but your reason and context for making notes are different. Keep asking questions and look for links to other references and modules. Notes should grab your attention, and make sense to you 10 weeks later. They may be longer or shorter than the original but paraphrased in your language. Just swapping a couple of words is not enough.

- **Diagrams** Using flow or spider diagrams as a first step in note making reduces the possibility of plagiarism. The style and language of the original author disappear behind a web of keywords and connecting arrows.
- **Check and share notes with friends** Everyone has different ideas about what is important, so comparing summary notes with a mate will expand your understanding.

6.3 Techniques

Note making is an activity where everyone has his or her own style. Aim to keep things simple, or you will take more time remembering your system than learning geography.

Which medium?

Hang on while I staple this note to my floppy disk.

Cards encourage you to condense material, or use small writing. Shuffle and re-sort them for essays, presentations and revision.

Loose-leaf paper lets you file pages at the relevant point and move pages around, which is especially useful when you find inter-module connections.

Notebooks keep everything together, but leave spaces to add new information, comments and make links. Index the pages so you can find bits!

PC's and *electronic organisers* allow notes to be typed straight to disk. This saves time later, especially with cutting and pasting references.

Multi-coloured highlighting

Saunders (1994) suggests a colour code to highlight different types of information on *your own* notes, books and photocopies. He suggests: 'yellow for key information and definitions; green for facts and figures worth learning; pink for principal ideas and links between things; blue for things you want to find out more about.' This approach requires lots of highlighter pens and consistency in their use, but it can be particularly useful when scanning your own documents. If this looks too complicated, use one highlighter pen – sparingly. On a pc, changing the font colour or shading will do the same task.

Coding

Coding notes assists in dissecting structure, and picking out essential points. During revision, the act of classifying your notes stimulates thoughts about the types and relative importance of information. At its simplest use a ** system in the margin:

****	Vital	**	Useful
?*	Possible	↓	A good idea but not for this, cross-reference to ...

A more complex margin system distinguishes different types of information:

| |	Main argument	B	Background or Introduction
|	Secondary argument	S	Summary
E.G.	Case Study	I	Irrelevant
[	Methodology, techniques	!!	Brilliant, must remember
R	Reservations – the 'Yes But' thoughts		
?	Not sure about this, need to look at ... to check it out.		

Opinions

Note your own thoughts and opinions as you work. These are vital, BUT make sure you know which are notes from sources, and which your own opinions and comments. You could use two pens, one for text notes and the other for personal comments. Ask yourself questions like: 'What does this mean?' 'Is this conclusion fully justified?', 'Do I agree with the inferences drawn?', 'What has the researcher proved?', 'What is s/he guessing?', 'How do these results fit with what we knew before?' or 'What are the implications for where we go next?'

Space

Leave spaces in notes, a wide margin or gaps, so there is room to add comments and opinions at another time. There is no time in lectures to pursue personal questions to a logical conclusion, but there is time when reviewing to re-focus thoughts.

Abbreviations

Use abbreviations in notes but not essays, intro. for introduction; omitting vowels Glc^{n} for glaciation; Hist^{l} for historical or using symbols. You probably have a system already, here are some sample abbreviations for faster writing.

+	And	=	is the same as
→	Leads to	xxx^n	xxion as in precipitation
↑	Increase	xxx^g	xxing as in pumping
↓	Decrease	//	Between
>	Greater than	Xpt	Except
<	Less than	←	Before
∴	Therefore	w/	With
?	Question	w/o	Without

Millions of unordered notes will take hours to create but not necessarily promote learning. Aim for notes which:

- are clear, lively, and limited in length;
- add knowledge and make connections to other material;
- include your own opinions and comments;
- are searching and questioning;
- guide or remind you what to do next.

Finally, feeling guilty because you haven't made some, or any, notes is a waste of time and energy.

6.5 References and Further Reading

Most student skills texts talk in detail about note making techniques.

Kneale, P.E. 1998 Notes for Geography Students, *Journal of Geography in Higher Education*, Directions, **22**, 3, 427–33.

Mills, C. 1994 Acknowledging Sources in Written Assignments, *Journal of Geography in Higher Education*, Directions, **18**, 2, 263–68.

Rowntree, D. 1988 *Learn How to Study: a guide for students of all ages*, (3rd edn) Warner Books, London.

Saunders, D. (ed.) 1994 *The Complete Student Handbook*, Blackwell, Oxford.

Lost the plot I

During field sampling we used a random stratified sampling design to investigate nine adjacent fields. Unfortunately, someone took some extra samples and added these points to the map. Then the field boundary lines were lost. Can you reconstruct the field pattern by drawing four straight lines only? Answer on p.254.

7 THINKING

I don't think I understood what I said either.

Few people set aside 'time to think'. Indeed, refusing to go out on the grounds that 'I have to stay in and think about the implications of positive discrimination as a factor in the self-perpetuating cycle of poverty in three developing cities,' will not boost your street cred. Stick with 'washing my hair' or 'I want a quiet night in with *EastEnders*.' For most people the effective stimulus to thinking is conversation and discussion. Being asked: 'What is your position on post-Fordism?' or 'How do you view the Chicago geographers advocacy of urban ecology as affecting our current understanding of interactions in the city?' can stimulate thoughts you didn't know you had.

Geography students are expected to apply their already well-developed thinking skills to a series of academic tasks and activities, to make reasoned judgements and arrive at conclusions about geographical issues. It is possible to pursue a geography degree at a rather superficial level, learning and re-presenting information. This is called surface learning. The aim of a university education is to practise the skills that move beyond this level to deeper learning, being active in questioning, relating ideas and opinions to other parts of the geography degree and to other subjects, and developing ones ability to inter-relate evidence and draw valid conclusions. This links to the ideas of deep study reading (see p.48).

A student's intellectual sophistication should mature during a degree course, but it is sometimes difficult to know what this might mean in practice. To convey some aspects of this development UTMU (1976) takes Bloom's (1956) list of cognitive skills for university students, and unpacks them by assigning a series of associated verbs (see Figure 7.1). A qualitative description of the anticipated development process for geography undergraduates at the

↓	**Knowledge**	Write; state; recall; recognise; select; reproduce; measure.
	Comprehension	Identify; illustrate; represent; formulate; explain; contrast.
	Application	Predict; select; assess; find; show; use; construct; compute.
	Analysis	Select; compare; separate; differentiate; contrast; breakdown.
	Synthesis	Summarise; argue; relate; precise; organise; generalise; conclude.
	Evaluation	Judge; evaluate; support; attack; avoid; select; recognise; criticise.

Figure 7:1 Bloom's (1958) skills (from UTMU, 1976)

University of Leeds, 1998, is outlined in Table 7.1. Most universities and many departments publish similar statements in university and department handbooks, start of year lectures and briefings. Think about where these statements match your experience. You are expected to progress from knowledge-dominated activities, to those with increased emphasis on analysis, synthesis, evaluation and creativity.

This chapter is a very minimal excursion into 'thinking' related activities. It is very brief and partial, ignoring most of philosophy and the cognitive sciences. It concentrates on three elements, unpacking what it means to be critical, reasoning, and questions to encourage and focus your thinking. If you feel your thinking activities could take a little more polishing then think through the ideas here. Like bicycle stunt riding, thinking gets better with time, not overnight. Thinking is tough.

7.1 Why do you think?

Thinking is used to acquire understanding and answers. Adjectives used to describe quality thinking include reasonable, clear, logical, precise, relevant, broad, rational, sound, sensible and creative. You cannot think in the abstract, there will be a purpose to your thinking, even if it is not immediately obvious. So the steps in quality thinking will involve:

- deciding on the objective (problem solving, understanding a concept);
- defining the background assumptions;
- acquiring data and information of a suitable standard to build up a reasoned argument;
- reasoning or inferring from the available information to draw logical conclusions;
- considering the consequences of the results.

How good you are at thinking is a matter for personal development and self-assessment. When tackling multi-dimensional geographical problems, make notes while thinking, plot your thoughts on spider diagrams, and record connections and links as they occur to you. Ideas float away all too easily.

7.2 Critical thinking

Critical thinking involves working through for oneself, afresh, a problem. This means starting by thinking about the nature of the problem, thinking through the issues

Level or Year	**Knowledge** *Broad knowledge and understanding of areas of geography. Fluency in subject vocabulary.*	**Analysis** *Problem-solving ability. Evidence of understanding. Ability to apply concepts to novel situations.*	**Synthesis** *Ability to bring together different facets of material, and to draw appropriate conclusions.*
1	Demonstrate a basic understanding of core subject areas, happy with geographical terminology. Demonstrate a knowledge of appropriate supporting analytical techniques (stats., pc, lab, and field).	Apply geographical techniques to real situations through class and field examples. Understand that there may be unique or multiple solutions to any issue. Appreciate the relative validity of results.	Be able to handle material that presents contrasting views on a topic and develop personal conclusions.
2	Demonstrate a comprehensive knowledge of specific subject areas. Be able to question the accuracy and completeness of information. Appreciate how different parts of the subject inter-relate.	Apply geographical theories to individual situations and critically examine the results. Understand that it may be appropriate to draw on multi-disciplinary approaches to analyse and solve geographical problems.	To locate and comment on diverse material, add personal research observations and integrate literature-based information with personal results.
3	Demonstrate a deep understanding of a number of specialist subject areas and methods. Appreciate the provisional state of knowledge in particular subject areas.	Understand how to solve problems with incomplete information, how to make appropriate assumptions, how to develop appropriate research hypotheses. Question and verify results.	Appreciate the breadth of information available. Identify and tap into key elements of the material. Produce coherent reports.
MA MSc	Demonstrate a broad, deep understanding of specialised subject areas and methods. Understand where this knowledge dovetails with the subject in general. Understand the current limits of knowledge.	Demonstrate ability to propose solutions to geographical problems involving appreciation of different approaches, gaps and contradictions in knowledge or data. Differentiation of unique and non-unique answers. Appreciation of reliability of a proposal or result given constraints and assumptions involved.	Be able to collate material from a wide range of appropriate geographical and non-geographical sources, integrate personal research material and collate the whole in a coherent, thoughtful and professionally appropriate manner. Be able to work to a specified brief.

Table 7.1 Skills matrix for geographers, The University of Leeds, 1998

Evaluation *Ability to review, assess and criticise one's own work and that of others in a fair and professional manner.*	**Creativity** *Ability to make an original, independent, personal contribution to the understanding of the subject.*	**Professionalism** *Ability to act as a practising geographer, to present arguments in a skilled and convincing manner and to work alone or in teams.*
Draw conclusions from results and identify the relative significance of a series of results. Evaluate the accuracy and reliability of information, results and conclusions.	Offer original comment on geographical material. Display or present information in different ways.	Be effective in planning and using time and geographical resources, including libraries and computer packages. Present information, written and orally, to a high standard.
Review existing literature and identify gaps, appraise the significance of results and conclusions.	Develop original, independent research skills, interpret data and offer comment. Be able to display information in a variety of ways.	Confident use of computer packages for analysis and presentation. Confident group worker and collaborator in research activities. Produce written work to a high professional standard.
Critically appraise information, evidence and conclusions from personal research and that of others.	Gather new information through personal research, draw personal conclusions and show where these insights link to the main subject areas.	Be able to set objectives, focus on priorities, plan and execute project work to deadlines. Produce well structured and well argued reports. Demonstrate fluency in oral and electronic communications.
Perform independent critical evaluation of information, evidence and conclusions, including their reliability, validity and significance. Be able to form and justify judgements in the light of contradictory information.	Offer insights into the material under discussion that are independent of data immediately available. Propose investigative approaches to geographical problems utilising geographical and non-geographical methods as appropriate.	Be able to make effective, confident presentations, answer detailed questions thoughtfully and clearly. Produce substantive reports that are well structured, well reasoned, well presented and clear. Work effectively as a team member and team leader.

and striving for a reasoned, logical outcome. During the process you need to be aware of other factors that impinge, where bias may be entering an argument, the evidence for and against the issues involved, and to search for links to other parts of geography. Essentially, critically evaluating the material throughout the process.

Being critical entails making judgements on the information you have at the time. It is important to remember that being critical does not necessarily imply being negative and derogatory. It also means being positive and supportive. It involves commenting in a thoughtful way. A balanced critique looks at the positive and negative aspects. Some students feel they cannot make such judgements because they are unqualified to do so. Recognise that neither you, nor your teachers, will ever know everything. You are making a judgement based on what you know now. In a year's time, with more information and experience, your views and values may alter, but that will be a subsequent judgement made in the light of different information.

Discussion is a major thinking aid, so talk about geography. It can be provocative and stimulating!

Where does *intellectual curiosity* fit into this picture? Geographical research is about being curious about geographical concepts and ideas. You can be curious in a general way, essentially pursuing ideas at random as they grab your imagination. We all do this. More disciplined thinking aims to give a framework for pursuing ideas in a logical manner and to back up ideas and statements with solid evidence in every case.

Uncritical, surface learning involves listening and noting from lectures and documents, committing this information to memory and regurgitating it in essays and examinations. The 'understanding' step is missing. Aim to be a deeper learner.

7.3 Reasoning

Strong essay and examination answers look at the geographical arguments, draw inferences and come to conclusions. Judgements need to be reasoned, balanced and supported. First, think about the difference between *reasoned* and *subjective reactions*, and reflect on how you go about thinking. Subjective reaction is the process of asserting facts, of making unsupported statements, whereas reasoning involves working out, or reasoning out, on the basis of evidence, a logical argument to support or disprove your case (see Table 7.2).

Create examples of reasoned rather than subjective statements with **Try This 7.1**. In your academic thinking and communications, avoid making emotional responses or appeals, assertions without evidence, subjective statements, analogies that are not parallel cases, and inferences based on little information unless you qualify the argument with caveats.

Subjective statement	Reasoned statement
'Ethnic minorities are a problem.'	'Smith (2010) shows that within inner city areas different ethnic groups raise different issues for social service provision. The cultural heritage and lifestyle patterns in contrasting groups means that the response to different sections within the community need to be appropriately tailored.'
'The Mississippi is a very dirty river.'	'Generally the Mississippi carries a very high sediment load during flood flows because high discharges erode bed and bank material and overlandflow entrains sediments from the extensive catchment. The sediment load will be radically reduced immediately downstream of any of the Mississippi reservoirs due to within reservoir sedimentation from the slower moving water.'

Table 7.2 Examples of reasoned statements

TRY THIS 7.1 – Reasoned statements

Either write a fuller, reasoned version of the four subjective statements below, OR pick a few sentences from a recent essay and rewrite them with more evidence, examples and references. Examples on p. 255.

1. The questionnaire results are right.
2. People are inflicting potentially catastrophic damage on the atmosphere and causing world-wide climate change.
3. The United States has become an urban country.
4. Pedestrianisation civilises cities.

7.4 Questions worth asking!

Being a critical thinker involves asking questions at all stages of every research activity. These questions should run in your head as you consider geographical issues. Questions like:

- What are the main ideas here?
- Are the questions being asked the right ones, are there more meaningful or more valid questions?
- What are the supporting ideas?
- What opposing evidence is available?
- Is the evidence strong enough to reach a conclusion?
- How do these ideas fit with those read elsewhere?
- What is assumed?
- Are the assumptions justified?
- What are the strengths and weaknesses of the arguments?
- Is a particular point of view or social or cultural perspective biasing the interpretation?
- Question definitions, even in seemingly 'objective' data sets.
- Is there evidence of high technical standards in analysis?
- Are the data of an appropriate quality?
- Do the results really support the conclusions?
- Are causes and effects clearly distinguished?
- Is this a personal opinion or an example of intuition?
- Have I really understood the evidence?
- Am I making woolly, over-general statements? *'There is a serious problem with housing in America'*. OK, but *'In the inner cites of the north eastern United States there is under-provision of basic housing'*, or *'The extensive development of retirement housing in Florida has taken a disproportionally high percentage of the state budget for 1992'* are better.
- Are the points made/results accurate? *'The River Wharfe has 2000 fish per 10 km.'* Sounds clear, but is it right? Where is the evidence?
- Are the results/points precise? *'The Ganges floods annually.'* This statement is clear and accurate, but we do not know how precise it is. Where does it flood?
- Is this information relevant? Keep thinking back to the original aims and argument. You can make statements that are clear, accurate and precise but if they are irrelevant, they do not help. Off the point arguments or examples distract and confuse the reader, and may lose you marks.

- Is the argument superficial? Have all the complexities of an issue been addressed? *'Should slums be swept away?'* A clear 'yes because …' will answer the question, but this is a very complex question requiring consideration of the economic, social, cultural, historical, political and planning perspectives. Ask what is meant by a slum, it has different connotations in Colchester and Calcutta.
- Is there a broad range of evidence? Does the answer take into account the range of possible perspectives? The question *'Discuss the arguments for and against nationalisation of the railways'*, could be argued from a UK Conservative Party economic point of view, but a broad essay might also consider Labour, Social Democrat, Green Party or Independent points of view with examples from five continents. You cannot cover all points in equal depth, but aim to make the reader aware that you appreciate there are other points or approaches.
- Are the arguments presented in a logical sequence? Check that thoughts and ideas are ordered into a sequence that tells the story in a logical and supported way.
- How does this idea or hypothesis fit with the wider field of enquiry? You might be looking at a paper on the *Incidence of asthma in Indonesian communities*, but where does it fit with medical geography, population geography or regional studies?
- Which examples will reinforce the idea?
- Can this idea be expressed in another way?
- What has been left out? Looking for 'gaps' is an important skill.
- Is this a definitive/true conclusion OR a probable/on the balance of evidence conclusion?
- What are the exceptions?

TRY THIS 7.2 – Gutting a paper

Select one paper from a reading list, any paper, any list! Make notes on the following:

Content:	What are the main points?
Evidence:	What is the support material? Is it valid?
Counter case:	What are the counter-arguments? Has the author considered the alternatives fully?
Summary:	Summarise relevant material from other sources that the author might have included but omitted.

How well did the author meet his or her stated objectives?

TRY THIS 7.3 – Comparing papers

Take three papers that are on the same or related topics, from any module reading list. Write a 1000-word review that compares and contrasts the contributions of the three authors. (Use the guidelines from the previous exercise.) Write 250 words on where these three papers fit with material from the module.

This seems like a major effort, but it really will improve your comprehension of a topic, so treat it as a learning exercise rather than an isolated skills exercise. Pick three papers you are going to read anyway. It is another approach to reading and noting.

Take a little time to think and reflect before jumping into a task with both feet. Having completed a task or activity, take a few minutes to reflect on the results or outcomes. **Try This 7.2** and **Try This 7.3** are two tutorial exercises which develop critical skills. Both provide frameworks for thinking, evaluating and synthesising geographical material.

In practice, few lecturers would argue that a logical perspective is the only way to deal with questions. Express your aesthetic opinions within an essay, ONLY if they are relevant and appropriate. An essay on 'the nature of flow processes in rivers' requires equations for fluid flow, Reynolds numbers and case examples. A poetic answer describing the delight you feel in viewing a flowing river at dawn will not do, and you will be in danger of being deemed 'a tributary short of a river'.

Can you improve the quality of your thinking on your own?

Yes, but it takes practice. You will probably become more disciplined in your thinking by discussing issues regularly. This is because the act of talking over an idea sparks off other ideas in your own mind. When someone else voices their point of view, you get an insight into other aspects of the problem. Thinking of arguments that run against your own position is difficult. A discussion group might do the following:

- ✔ start by summarising the problem;
- ✔ sort out objectives to follow through;
- ✔ share data and evidence, the knowledge element;
- ✔ share views on the data, 'I think it means ... because ...';
- ✔ work out and discuss the assumptions the data and evidence is making;
- ✔ discuss possible implications; evaluate their strengths and weaknesses;
- ✔ summarise the outcomes.

A good reasoner is like a good footballer; and becomes more adept by practising.

Where to think?

Thoughts and ideas arrive unexpectedly and drift off just as fast unless you note them. Take a minute to recall where you do *your* thinking. There are almost as many varied answers as people, but a non-random sample of individuals in a lecture (N = 67) shows favoured locations include in bed at 4 a.m., while walking to work, jogging, swimming, working out in the gym, cleaning the house or cutting grass! There is certainly a common element of thinking being productive while otherwise engaged in an activity that allows the mind to wander in all sorts of directions without distractions like phones and conversations. The majority of students who offered 'walking' and 'the gym' as their best thinking opportunities are evidence of this. Writing down ideas is vitally important, but is incompatible with aerobics. Recognise this problem by taking 10 minutes over a drink after exercise, or a couple of minutes at a bus stop, to jot down thoughts and plans. This makes aerobic exercise an effective thinking, multi-tasking activity.

Avoiding plagiarism

Good thinking habits can minimise your chance of inadvertently plagiarising the work of others. Get into the habit of engaging and applying concepts and ideas, not just describing or reporting them. That means thinking over the ideas to find your own contexts and alternative examples. Make sure you include your own thoughts, opinions and reflections in your writing. Be prepared to draft and redraft so that the thoughts are in your own language, and acknowledge your sources. Leave time to link ideas coherently. Finally, put the full reference for your citations at the end of each piece of written work.

Thinking and understanding involve a commentary in your head. Writing a summary in your own words is a good way to check you understand complex ideas. Ask:

'Do I understand this?' at the end of a page, chapter, paper, tutorial, lecture … and not just at the end.

7.5 References and further reading

Bloom, B.S. (ed.) 1956 *Taxonomy of Educational Objectives: 1 Cognitive Domain*, Longman, London.

Buzan, T. and Buzan, B. 1993 *The Mind Map Book*, BBC Books, London.

UTMU 1976 *Improving Teaching in Higher Education*, University Teaching Methods Unit, London.

Van den Brink-Budgen, R. 1996 *Critical Thinking for Students: how to use your recommended texts on a college or university course*, How To Books, Plymouth.

Geo-quick crossword I

Answers on p.255

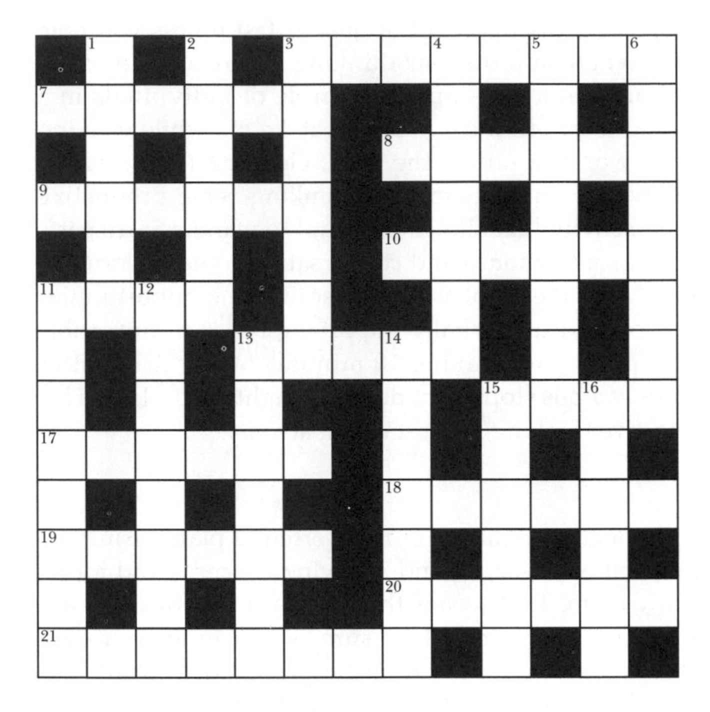

Across

3 Tossed away? (3-5)
7 Lakeside tide (6)
8 Tree lizard (6)
9 Group of friends (6)
10 A single Gurkha? (6)
11 Sales point (4)
13, 15 On the brink at Beachy (5, 4)
15 (see 13 ac.)
17 Papal Ambassador (6)
18 Equal pressure line (6)
19 Noah's mooring point (6)
20 A young 16 (6)
21 Traders' storehouse (8)

Down

1 Acme or culmination (6)
2 The coldest hat? (3-3)
3 Free lad (anag) (6)
4 Minimal botanical cover (3, 4)
5 Not reduced (8)
6 Agile cat (anag) (8)
11 An early, rocky, culture (5, 3)
12 Surface excavation (4, 4)
13 Elemental conditions (7)
14 Ship load, or a load to ship (7)
15 Glamorous alien (6)
16 He browses around (6)

8 CONSTRUCTING AN ARGUMENT

Two geographers walking through the city observed two people shouting across the street at each other. The first said, 'Of course they will never come to an agreement'. 'And why?' enquired his mate. 'Because they are arguing from different premises.'

At the centre of every quality geographical essay, report or presentation is a well-structured argument. This is an argument in the sense of a fully supported and referenced explanation of an issue. In producing an argument you are not looking to provide a mathematical proof, or evidence that is strong enough to support a legal case. You are attempting to establish a series of facts and connections that link them, so that your deductions and conclusions appear probable. If you establish enough links and supply the case evidence through real geographical examples, your argument cannot be rejected as improbable.

Start by thinking about your motive for putting pen to paper, not just the 'I want a 2.1', what do you want to say? Most geographical issues can be viewed from a number of aspects, and exemplified with a wide range of material. This means unpacking the elements of an argument and structuring it logically. Your aim might be:

- to develop a point of view as in a broadsheet newspaper style editorial or a debate;
- to persuade the reader, possibly your geography examiner, that you know about 'post colonial changes in agriculture in the developing world' or 'the role of spatial modelling to forecast the spread of disease'.

You may be more specific, perhaps wanting to define:

- how one concept differs from the previous concepts of . . . ;
- the implications of using one idea or concept rather than another one;
- whether a decision is possible or 'Are we still sitting on the fence?';
- what would need to be adapted, amended, what would work and what would not, if an idea or concept was transferred to a new environment.

Some arguments are linear, others rely on an accumulation of diverse threads of evidence which, collectively, support a particular position. Recognise that different geographers will present equally valid but conflicting views and opinions. In considering rural transport issues you could seek information from an ecologist, agricultural scientist, transport expert, political scientist, local councillor, employment experts, environmental campaigning groups, heritage protection organisations, local people, and bus and car drivers using the routes in question. In developing a statement about rural transport you would need to present the different views and consider:

- What are the limitations of each view?
- What elements are entrenched?
- How might a consensus be formed?

A geography research project can leave you overwhelmed with evidence. 'Thinking' is involved in designing an approach to maximise your understanding of influences and interactions between different facets and elements of the data and presenting these in a coherent argument. Points of view should be part of the chain in an argument, or arise from the argument. The weight of the evidence or ideas makes the point of view acceptable.

8.1 Structuring arguments

Any argument needs to be structured (*see* Figure 8.1) with reasoned evidence supporting the statements. A stronger, more balanced argument is made when examples against the general tenet are quoted. Russell (1993) describes three classic structures that can be adopted in arguing a case (see Figure 8.2). It is this author's contention that at University level the third model should be used every time, and that in every discussion, oral presentation, essay, report and dissertation you should be able to point to each of the six sections and to the links between them.

8.2 Unpacking arguments

It is all too easy when speaking and writing to put too much information into a sentence, or make very general statements. One might say 'Poverty in Africa is a consequence of the modern world system'. This is true but hides much information. A fuller statement like 'Inequality in Africa has complicated historical origins. The roots of poverty in contemporary Africa can be traced back to earlier historical periods, particularly colonial times, and can be seen as a consequence of the unequal relations that exist between "North" and "South". It has also been argued that inequality in Africa is a consequence of the emergence of a capitalist world economy.' To further strengthen this example add references as in: 'a capitalist world economy (Dollar and Pound, 2010)'.

Another statement meriting some unpacking is 'Men and women play different social roles in developing countries.' A more detailed statement might say

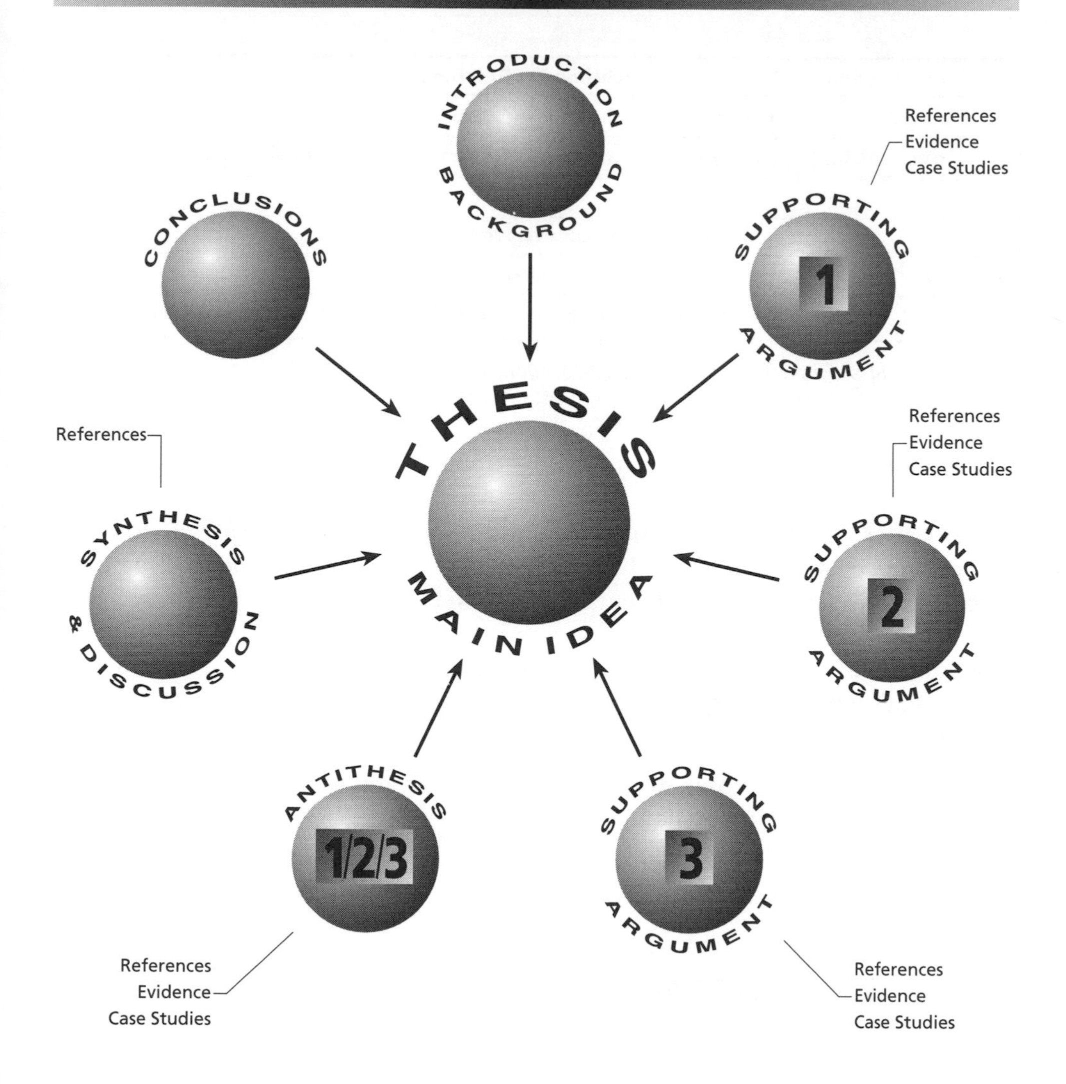

Figure 8:1 A spider diagram template for an argument

'Gender is an important factor in the practice and experience of development. In many societies men and women have clearly defined and often distinct roles. The different tasks assigned to men and women in each society comprise an important part of their gender identities.'

From the soils area 'Microbiological processes control plant nutrient uptake.' This is a straightforward sentence and might be an essay title, just add Discuss. A version, gaining more marks by adding detail, might read:

1	**Introduction**	**Thesis**	**Conclusion**	**References**		
2	**Introduction**	**Thesis**	**Antithesis**	**Conclusion**	**References**	
3	**Introduction**	**Thesis**	**Antithesis**	**Synthesis**	**Conclusion**	**References**

Thesis	The argument under consideration
Antithesis	The argument against the thesis
Synthesis	The balancing of the different points of view made in the thesis and antithesis

Figure 8:2 Classic models for presenting argument (after Russell, 1993)

> Soil micro-organisms, which include bacteria, mycorrhizal fungi, actinomycetes, protozoa and algae control a number of soil processes, recycling and releasing nutrients for plants and other soil organisms. The presence or absence of oxygen determines the species present and their activity rates. The mycorrhizal fungi for example, are symbiotic dwellers on host plant tissues, taking C from the plant and making P and other nutrients available for plant uptake.

When reviewing your own writing, identify 'general' sentences which could benefit from a fuller explanation.

8.3 Rational and non-rational arguments

Arguments are categorised as being rational and non-rational. The non-rational are to be avoided wherever possible. Here are five examples:

1. Avoid bald statements like 'Rio de Janeiro has the worst/best/smallest/greatest ... slums/markets/housing policies.' This is hyperbole. It might sell newspapers but it is not an argument unless supported by data.
2. 'The principal problem for refugees is absence of income. Students have no income therefore students are refugees.' This is a very poor argument and untrue in the majority of cases. Some students are refugees, most are not.
3. Going OTT with language or throwing in jargon to impress the reader. This is a typical journalistic device, involving an emotional rather than factual appeal to the reader. 'Inner city housing in Thriftston was run-down, the area generally

subject to neglect and depopulation' is a more considered academic statement than 'Thriftston: an abandoned, riot-torn, rubble-strewn, neglected and deserted city' unless of course the situation is extreme. Describing Sarejevo in 1997 in this way would be OK, for central Sheffield it would be somewhat biased.

4. 'It is unquestionably clear that bacteria control all soil processes.' Words which sound very strong like clearly, manifestly, undoubtedly, all, naturally, and obviously, will influence the reader into thinking the rest of the statement must be true. Overuse of strong words is unhelpful, the written equivalent of browbeating or shouting, used sparingly they have greater effect.
5. 'All Third World cities have sanitation problems' or 'All penguins are found in the Antarctic'. The global scope of geography means that an exception can be found to almost every generalisation.

Inductive or deductive? Which approach?

In presenting an argument, orally or on paper, give some thought to the ordering of ideas. If you write deductively you begin with the general idea and then follow on with examples. Inductive writing starts with specific evidence and uses it to draw general conclusions and explanations. You must decide what suits your material. In general, use the inductive approach when you want to draw a conclusion. The deductive approach is useful when you want to understand cause and effect, test an hypothesis or solve a problem. The examples given here of an inductive (A) and deductive (B) approach are in paragraph form, but the same principles apply to essays and dissertations. There is an enormous leap here from 23 patients to a generalisation about health care in the developing world, it would benefit from further unpacking with more examples.

A In a snapshot survey of 23 patients in a hospital in Papua New Guinea, Clark (1998) reported that recovery was inhibited by poor nutrition, infection and the late presentation of wounds for care after unsuccessful 'bush' remedies. The local diet is carbohydrate rich, dependent on sago, taro and sweet potatoes with little protein from meat or fresh fruit. Infection was high due to poor hygiene, it is difficult to keep wounds clean while continuing with field work, and ulcers form easily. In hospital, drugs are limited in type and quality, and staff are few in number and cannot be expert in all medical sub-disciplines. Although hospital resources are limited, it must be recognised that they are already supplemented by overseas and charitable aid. The government contribution to healthcare is about 50 per cent. While this situation continues in developing countries like Papua New Guinea, medical care will be characterised by limited healthcare options and effectiveness.

B Developing countries have limited funds for medical resources, which restricts health care options and effectiveness. While other factors may affect patient behaviour and recovery, under-funding is an important issue. Approximately 50 per cent of health care in Papua New Guinea (PNG) is provided by the government, the remainder is funded by overseas aid through Christian missionary organisations. Clark (1998) in a limited study of patients at a hospital in PNG receiving valuable overseas funding, discussed factors affecting wound

> recovery. Poor nutrition and high wound infection rates exacerbate the problem. Within the hospital there is a limited selection of drugs, and blood supplies. Staff expertise is necessarily limited when there is only one surgeon to deal with all cases. Nurses can treat less serious cases, but this is not ideal. Despite staff enthusiasm and best practice given the circumstances, lack of medical resources limits their effectiveness.

Incidentally, the Clark (1998) reference is from a medical journal. Look for supporting case studies in soils, geology, agriculture, politics, economics ... journals, as appropriate for your study, not just the geographical literature.

8.4 Potential minefields

Watch out for bad arguments creeping into your work. The following examples are rather obvious. Be critical in reviewing your own work, are there more subtle logic problems? It is easy to spot logical errors when sentences are adjacent, they may be less easy to see when there are over a thousand words of an essay in-between.

✗ **Circular reasoning:** Check that conclusions are not just a restatement of your original premise.

Urbanisation is a continuous process. We know this because we can see it going on around us all the time.

✗ **Cause and effect:** Be certain that the cause really is driving the effect.

Students are forced to live in crime-ridden areas, so crime becomes part of student life.

✗ **Leaping to conclusions:** The conclusion may be right, but steps in the argument are missed, see unpacking arguments (section 8.2). Some arguments are simply wrong!

Our survey of five river reaches showed sediment grain size ranged from 0.0002–10mm in diameter. We therefore concluded that boulders were not a feature of this river.

Division of labour assigns specific production processes to individuals in the workforce. This leads to class divisions in society.

Be clear about the difference between arguments where the supporting material provides clear, strong evidence, and those where there is statistical or experimental evidence supporting the case, but uncertainty remains. In student projects and dissertations, time often limits experimental or field work. There should be fifty samples, but, tragically, there was only time to get six. Be clear about the limitations. Statements like 'On the basis of the six samples analysed we can suggest that ...' or 'The statistical evidence suggests ... However, the inferences that may be drawn are limited because sampling occurred over one summer and at three sites in Kowloon' are very acceptable. Qualifying statements of this type have the additional merit of implying that you have thought about the limits and drawbacks inherent in the research results.

Is this a good argument?

Watch out for arguments where the author gives a true premise, but the conclusion is dodgy. Just because you agree with the first part of a sentence does not mean that the second part is also right. Keep thinking right through to the end of the sentence. Having articulated an argument; do you buy it? Why? Why not? What is your view? Look at **Try This 8.1** as a starter.

TRY THIS 8.1 – Logical arguments?

Consider the following statements. What questions do they raise? Are they true and logical? What arguments could be amassed to support or refute them? See p.255 for some responses.

1. Divergent plate margins occur where two plates are moving away from each other, causing sea-floor spreading.
2. Rainforests store more carbon in their plant tissue than any other vegetation type. Burning forests release this stored carbon into the atmosphere as CO_2. The net result is increased CO_2 in the atmosphere.
3. If the system produces a net financial gain, then the management regime is successful, and the development economically viable.
4. Urban management in the nineteenth century aimed to reduce chaos in the streets from paving, lighting, refuse removal and drainage, and to impose law and order.

Language that persuades

Strong words often connote weak arguments.

If you use strong statements like 'Clearly we have demonstrated …', 'This essay has proved….' or 'Unquestionably the evidence has shown….', be sure that what you have written really justifies the hype. It is worth thinking about your stock of linking phrases, use **Try This 8.2** to review what you currently use, and then look at the answers to see if there are others to add.

- When reviewing your work read the introduction and then the concluding paragraph. Are they logically linked?
- Keep the brain engaged.

TRY THIS 8.2 – Logical linking phrases

Identify words and phrases that authors use to link arguments by scanning through whatever you are reading at present. Some examples are given on p.256.

8.5 Opinions and facts

It is important in both reading and writing to make a clear distinction between facts supported by evidence and opinions, which may or may not be supported. Telling them apart is a matter of practice in looking to see what evidence is offered and what else might have been said. Has the author omitted counter-arguments? In **Try This 8.3** there are some short extracts to practise your discrimination skills, which are facts, which opinions, and which have evidence offered in support.

TRY THIS 8.3 – Fact or opinion

Are these statements factual or based on opinion? Briefly describe the argument advanced by the writer. (All statements are taken verbatim from articles and reviews in the *Transactions of the Institute of British Geographers*, **22**, 4, 1997, see page numbers to follow up.) Suggested answers are on p.256.

The city has become 'trendy.' (Jencks 1996). (p.411)

Urban studies has experienced a remarkable renaissance in the past fifteen years, fuelled by the replacement of tight, positivistic approaches with structuralist and, more recently, post-structuralist theories. A veritable deluge of newspaper and magazine reports now addresses urban crises and 'regeneration' processes. (p.411)

Dam construction has altered flooding patterns in most major river basins in tropical Africa, for example on the Tana River in Kenya or the River Benue in Cameroon downstream of the Lagdo Barrage (Drijver and Marchand, 1985). (p.430)

Definitions of 'sustainable development' abound. (p.431)

Hansen has described the importance of mountain climbing to the professional middle classes of late nineteenth-century Britain. They 'actively constructed an assertive masculinity to uphold their imagined sense of British imperial power' (Hansen, 1995, 304). (p.450)

Women were unsuited by 'sex and training' for 'exploration' and, since geography was not about library work, like the Asiatic Society, or merely a social club, like the Zoological Society, then, to preserve its focus on exploration, it had to exclude women from its Fellowship (*The Times* 31 May 1893, 11d). (p.457)

This analysis has documented changes in regulatory practice and the implementation of food policy during a period of increased salience of food quality concerns on the one hand and the intense competitive efforts of corporate retailers to maintain and develop their markets on the other. (p.485)

The transcriptions of the (40) interviews have been woven into and inform an account of a complex set of (partial and uneven) transitions in the political, cultural and what might best be called the 'moral economy' within which the discipline of geography is reproduced. (p.488)

A major problem with **Try This 8.3** is that only part of a paragraph is reproduced, and inevitably some opinions are based on limited information. As in life, if you start looking for arguments you will find them. **Try This 8.4** is a way of getting into analysing arguments. Use it as a framework next time you read something geographical.

TRY THIS 8.4 – Spot the argument

As you read any geographical article: (a) highlight the arguments; (b) highlight the statements that support the argument; and (c) highlight the statements that counteract the main arguments. Is the information presented balanced?

8.6 Unfashionable arguments

Examining all sides of a question can be particularly difficult if moral, unfashionably moral or ethical elements are involved. Consider how you would discuss the causes of urban rioting. You might say that people riot because it is fun, a group reaction, an opportunity to take personal revenge, a way of livening up a dull, hot evening, an overreaction to a minor misunderstanding that escalated beyond control and reason. You could talk about the 'fact' that some people are 'evil', not 'made evil by circumstance' and state that behaviour is not solely determined by upbringing, school, affluence and employment prospects. A more pc (politically correct) or 'right on' answer might ignore these elements and discuss social deprivation, unemployment, police brutality, sub-standard housing, (mis)use of drugs and racial tension. The rioters may be portrayed as victims of circumstance, 'it's not their fault', rather than the active participants, inciters, throwers of bricks and bottles; people responsible for their own actions.

These issues are, of course, enormously difficult. Inter-locked, inter-linked and coloured in personal understanding by background and culture. The challenge for you, as an unbiased (is this possible?) reporter, is to strive to see all perspectives, including the 'moral' and the unfashionable. If the issue is 'debt', is it because 'credit' is too easily given, personally or nationally? If shoplifting is an economic problem, is it because 'shops make goods too accessible to the customer'?

One entry point might be to think of the headlines that will never make the papers:

NO RIOTING IN LOS ANGELES/TOXTETH/BRISTOL ON 36518 DAYS THIS CENTURY

Why have there been no riots in most cities on most nights of this century?

8.7 Examples of Arguments

It is impractical in this text to reproduce an essay and discuss the quality of the arguments, rather, some students' answers, together with some examiners' comments are used to make points about the adequacy of the arguments. This section could also be useful when revising for short answer examination questions.

As an exercise in exemplified brevity, expressing ideas or concepts in two or three sentences, or a maximum of 50 words is a good game. Consider these three answers to the question:

1. Define feedback. (2 sentences maximum)

A. *Feedback occurs where part or all of the output from one process is also the input for another process. This can be positive, negative or both.* (26 words)

B. *This is where the system effectively alters itself. Positive feedback is where the signal is reinforced, the converse is called negative feedback.* (22 words)

C. *Feedback is the action by which an output signal from a process is coupled with an input signal (Figure 8:3) Feedback may be positive, which reinforces any changes, e.g. the greenhouse effect or change in albedo, or negative which ameliorates any changes.* (49 words including the diagram and its caption.)

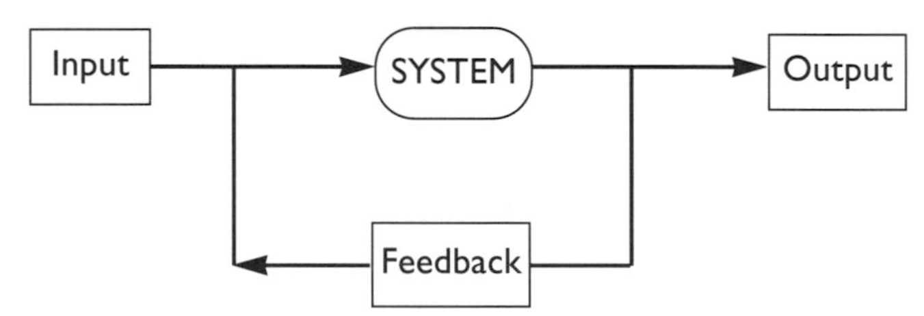

Figure 8:3 A simple feedback loop

These are not perfect answers, but the third statement has a little more geography and the diagram reinforces the argument. This, and the next question, ask for a supported, factual response rather than an argument.

2. What is albedo?

A. *Albedo is the ratio of the total incident reflected electromagnetic energy to the total incident electromagnetic energy falling on a body. Albedo, expressed either as a percentage or on a 0–1 scale, varies between surfaces, and generally increases with roughness. The cryosphere has a high albedo 70–80%, whilst the tropical rainforest has a relatively low value of 20–30%.*

B. $$\textit{Albedo} = \frac{\textit{Total Radiation Reflected}}{\textit{Total Incoming Radiation}}$$

It is measured on a 0–1 scale where 0 = low, 1 = high. The average albedo of the earth is 0.3.

I think both these definitions would get full marks since both define albedo, but if there are bonus marks going, or within an essay, the first answer has some added value. Using the equation format perhaps makes the second answer clearer than the first. Inserting equations in essays is fine. Both answers have indicated the units involved.

The next three examples build up arguments with evidence.

3. Explain the steps involved in deduction in scientific geography.

A. *Deduction is where a relationship is deduced from geographical phenomena, whereas induction looks for relationships within data or information. In deduction a general idea, such as 'more people live in towns', is formulated as a hypothesis that may be tested as in: 'population densities increase as you move from rural areas to city centres'. Having defined the hypothesis it can then be tested in a scientific manner in order to verify or disprove it. Tests might involve primary data collection through questionnaires, observation, or analysis of census and ward data or local authority data. Following analysis the hypothesis will be accepted, rejected or accepted with caveats. Appropriate deductions might involve a statement like 'The hypothesis was shown to be generally true for Bloggsville, with two exceptions. The 1930s' peripheral council housing estates and areas with recent, 1990s', greenbelt housing developments (Figure x) have high population densities but the general model forecasts lower densities.'*

The first sentence is a little bald, (is it right?) but indicates the writer is aware of an alternative, inductive approach. The remaining sentences have integrated examples. I hope you will agree that this answer gets more marks than the following bullet-style answer:

B. *The steps in deductive reasoning are: have an idea; develop a working hypothesis; test the hypothesis experimentally through data collection and analysis; decide whether to accept or reject the hypothesis and further explain how the world works in this case, and then reassess or redefine the problem if the research is to continue.*

This short statement answers the question at a level that will get a pass mark, but it has no geography content. The first answer will score additional marks if the lecture or main text example is not to do with housing and distance decay.

4. Why should spatial models be linked within GIS packages?

Current GIS models can manipulate large quantities of data speedily and accurately, and present it visually at a very high standard, through buffer and overlay techniques. Display functions are important and can be very useful in explaining complex relationships to non-specialists, but the standard in display has yet to be matched in analysis capability. GIS packages currently lack sophisticated geographical data analysis functions and have to be coupled to other programmes for detailed modelling of traffic flows, global temperature change or erosion forecasting. For GIS packages to be useful research analysis tools in planning or model building, they need to move beyond their current rather simplistic geodemographic and location functions to incorporate models that could look, for example, at competition, interaction and gravity modelling. As spatial interaction modelling is incorporated into GIS systems, it is vital to monitor the accuracy of model forecasts and to compare this with traditional spatial modelling techniques to evaluate the relative merits of the two systems.

This question looks for an argument, 'why' is a keyword. The author explains what current GIS models can do and shows where linking to spatial models would be advantageous. The first sentence could have a stop after the second comma, but is strengthened by the two examples. The section *'detailed modelling of traffic flows, global temperature change or erosion forecasting'* is good but would be further strengthened by including references (REF.) after each of the three examples, *'traffic flows (REF.), global temperature change (REF.) or erosion forecasting (REF.)'*. The final sentence is a neat caveat that shows the author knows that just adding functions to packages does not necessarily improve forecasting quality.

5. Discuss the value of a structured interview compared with a postal questionnaire.

A structured interview has the benefits of being more personal, the interviewer gets a 'hands on' feel for the data, and problems can be detected and dealt with as the interviews progress, for example by extending, adding or dropping some questions. The interviewee should feel involved, which is likely to encourage full answers and s/he may volunteer additional information. There is a problem with the honesty of answers in all survey research, but a one-to-one interview is more likely to elicit honest responses.The number of interview refusals is likely to be small. Conversely, interviews are very time-consuming in both administration and analysis, especially if there are audiotapes to interrogate.

A postal questionnaire can feel very impersonal, there is little motivation to complete, (even Reader's Digest *offers wear thin). Since the number of returns may be as low as 15 per cent, it must be sent to a large population to account for non-returns. Some respondents may misunderstand questions, so instructions must be clear and questions unambiguous.*

Both approaches require careful preparation of questions, with questions structured in a user-friendly order. A pilot survey should be carried out in each case and the schedule adjusted in response to the results.

If we are looking at argument in isolation then this is good. BUT, this answer would be as useful in a sociology or psychology essay and gain the same marks. It has argument, structure and lots of good points but no links to geography. No geographical examples and no references to the many authors who have written about social survey techniques in geography. In a geography examination it would score 45–50 per cent, a pass, but it needs the geographical angle to do better.

Use short sentences

- Avoid 'cop out' statements like 'Others disagree that. ...' Who are these others????? There are more marks for 'Knowitall (2010) and Jumptoit (2010) disagree making the point that. ...'
- No examiner will give marks for the use of '...etc.'.

Yes but ...

In addition to logical linking phrases keep a list of caveat or 'yes but' statements handy. There are alternatives to 'however' like consequently; as a result; by contrast; thus; albeit; therefore; so; hence; nonetheless; despite the fact that; although it has been shown that. More extended versions include: 'A rigorous qualitative geographer might argue that the results of this research are ethereal, confused and disorganised, and that some structure would have helped the project'; 'The author gives an interesting but superficial account of ...'; 'The figures show ... but were not able to support or refute the main hypothesis because ...'; 'The succeeding tests support this criticism because ...'; 'The outcome may be influenced by ...'; 'Only a few of the conclusions are substantiated by the experimental analyses'; 'The argument is stated but the supporting evidence is not given'; 'Although this is an entirely reasonable exploratory approach it neglects ..., and ... thereby weakening the inferences that may be drawn'; 'Therefore, the criticism should be directed at ... rather than at ...'; 'The outcomes, therefore, relate to a different set of conditions to those initially outlined.' Build on this list as you read. As you research use these 'yes but ...' phrases to focus thinking and to draw valid and reasoned inferences.

Be critical (within reason) of your writing and thinking. This means allocating time to read critically, remove clichés and jargon, and add caveats and additional evidence.

Top Tips

- Write in clear sentences, with geographical cases and references to support the statements.
- Give due weight to arguments that support and refute your main argument.
- Talk through your arguments with friends. Explaining and persuading someone of your case usually clarifies arguments.
- Ask 'Does this persuade me?'
- Use Figure 8.1 as a template to check that you have a balanced argument.

8.8 References and further reading

Clark, M.P.A. 1998 Wounds of Wewak, Papua New Guinea, *Journal of the Royal College of Surgeons of Edinburgh*, **43**, 3, 174–7.

Fairburn, G.J. and Winch, C. 1996 *Reading, Writing and Reasoning: a guide for students*, (2nd edn), Open University Press, Milton Keynes.

Russell, S. 1993 *Grammar, Structure and Style*, Oxford University Press, Oxford.
Toulmin, S., Rieke, R. and Janik, A. 1979 *An Introduction to Reasoning*, MacMillan, New York.

Geo-links ladder and town ladder

Change one letter at a time to make new words, each time you move down the ladder. Answers on p.258

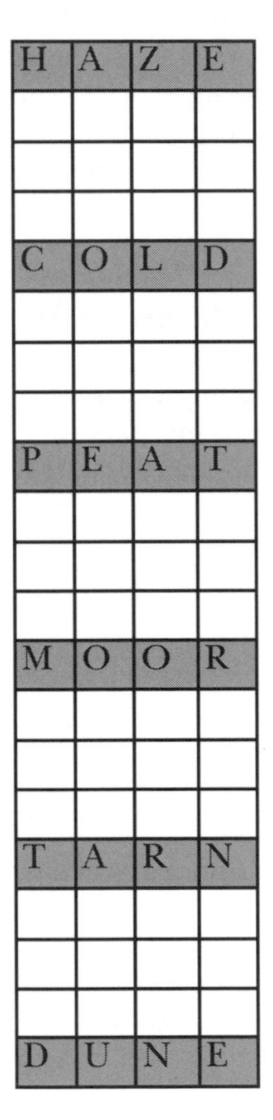

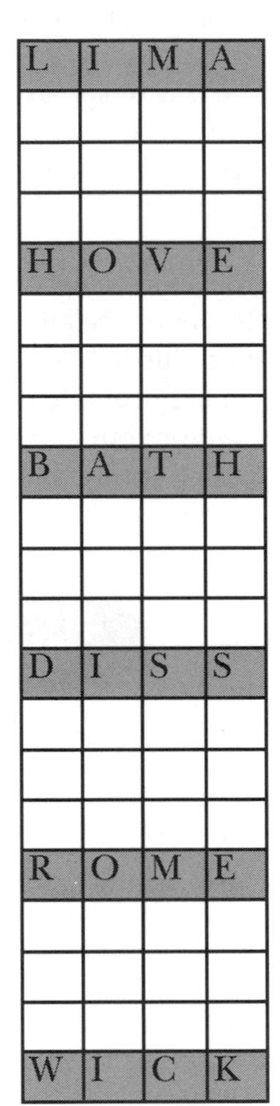

Geograms 2

Try these geographical anagrams. Answers on p.258.

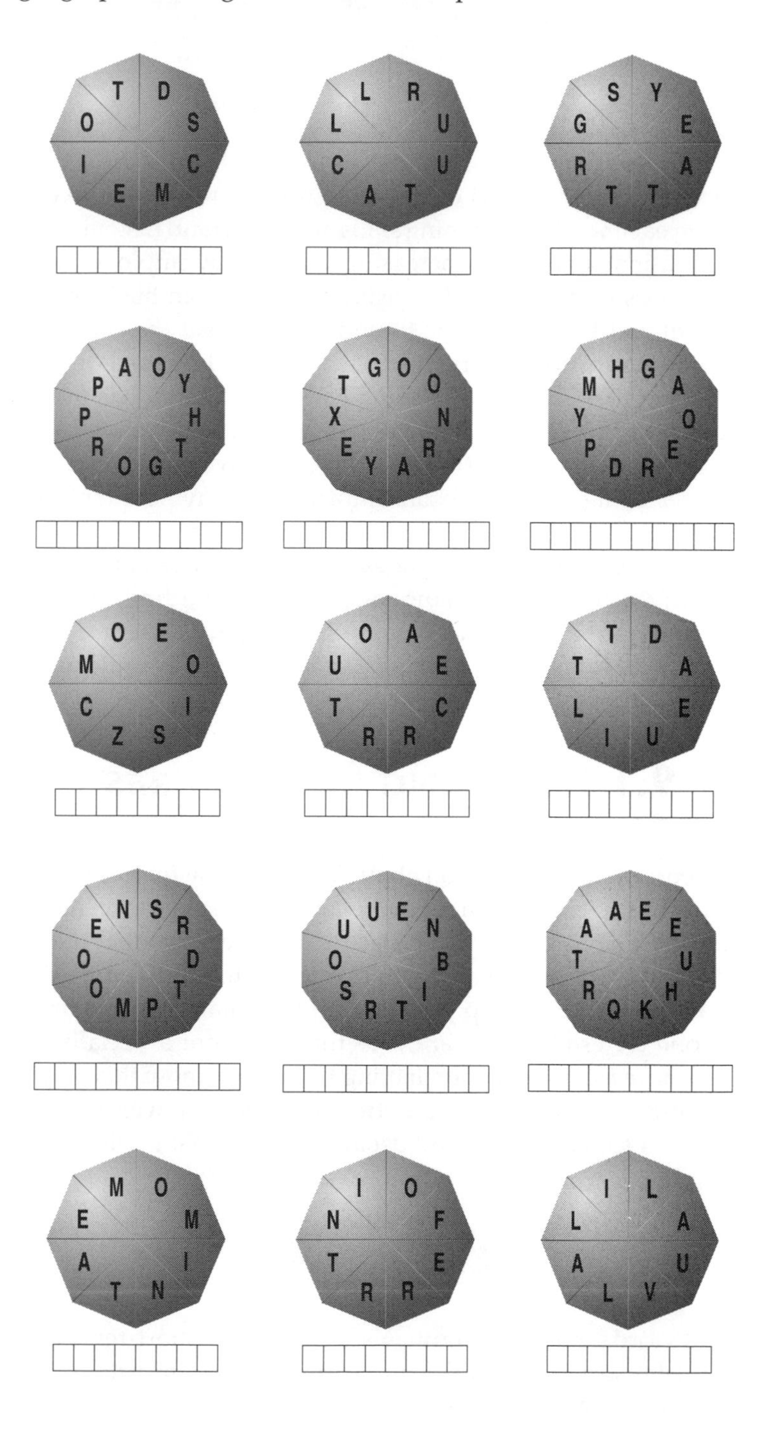

9 LISTENING

Hearing is easy, listening is tough.

Geography lectures, seminars and discussion groups can involve twelve or more hours of listening each week, so listening skills are important. Listening is often considered an automatic activity, but is increasingly quoted by employers as a vital skill for effective business performance (Gushgari *et al.* 1997). In business, valuing the customer or client and taking the time to understand what the speaker is conveying, gives a better impression. Interviewing is a standard research methodology in geography (Lindsay 1997). Therefore, being good at listening has the potential to enhance the quality of your research understanding and inferences.

Listening is not the same as hearing, it is a more active and inter-active process. Listening involves being ready to absorb information, paying attention to details, and the capacity to catalogue and interpret the information. In addition to the actual material and support cues like slides and OHTs, there is information in the speaker's tone of voice and body language. As with reading, the greater your geographical background, the more you are likely to understand. Which means that listening is a skill involving some preparation, a bit like parachuting.

9.1 LISTENING IN LECTURES

Arriving at a lecture with information about last night's activities or juicy scandal is normal, but the brain is not prepared for advanced information on medieval housing or the Loch Lomond readvance. Some lecturers understand that the average student audience needs 5 minutes background briefing to get the majority of brains engaged and on track. Others leap in with vital information in the first 5 minutes because 'everyone is fresh!' Whatever the lecture style, but especially with the latter, you will get more from the session having thought 'I know this will be an interesting lecture about …' and scanned notes from the last session or library. Assuming from the start that a lecture will be dull usually ensures that it will seem dull.

Top Tips

- A lecturer's words, no matter how wise, enter your short-term memory, and unless you play around with them and process the information into ideas, making personal connections, the words will drop out of short-term memory into a black hole. Think about the content and implications as the lecture progresses.

- You may feel a lecturer is wildly off beam, making statements you disagree with, but do not decide he or she is automatically wrong, check it out. There might be dissertation possibilities.
- Keep a record of a speaker's main points.
- Be prepared for the unpredictable. Some speakers indicate what they intend to cover in a lecture, others whiz off in different directions. This unpredictability can keep you alert!, but if you get thoroughly lost, then ask a clarifying question (mentally or physically), rather than 'dropping out' for the rest of the session.
- If you feel your brain drifting off ask questions like 'what is s/he trying to say?' and 'where does this fit with what I know?'
- Have another look at p.5, expectations of lectures.

Treat listening as a challenging mental task.

9.2 Listening in discussions

No one listens until you make a mistake.

Discussion is the time to harvest the ideas of others. With most topics in political, social, economic and cultural geography there are such a diversity of points of view that open discussion is vital. Endeavour to be open-minded in looking for and evaluating statements which may express very different views and beliefs from your own. Because ideas fly around fast, make sure you note the main points and supporting evidence (arguments) where possible. Post-discussion note collation is crucial, mostly involving ordering thoughts and checking arguments that support or confute the points. Have a look at **Try This 9.1** and think about how you score on effective listening.

9.3 Telephone listening

Some research interviews must be conducted over the telephone, and companies are increasingly conducting first job interviews on the telephone. Listening and talking under these circumstances are difficult to do well. If video-conferencing

TRY THIS 9.1 – Assess your listening effectiveness in discussions

Think back over a tutorial or a recent conversation and rate and comment on each of these points. Alternatively score this for a friend or fellow tutee, and then think about your skills in listening, compared with theirs.

Score your effectiveness on a 1–4 scale, where 1 is No, and 4 is Yes		**Comment**
Did you feel relaxed and comfortable?	1 2 3 4	
Did you make eye contact with speaker?	1 2 3 4	
Were you making notes of main points and personal thoughts during the discussion?	1 2 3 4	
Did you discuss the issues?	1 2 3 4	
Were you thinking about what to say next while the other person was still talking?	1 2 3 4	
Did you ask a question?	1 2 3 4	
Did you get a fair share of the speaking time?	1 2 3 4	
Did you empathise with the speaker?	1 2 3 4	
Did you accept what the others said without comment?	1 2 3 4	
Did you interrupt people before they had finished talking?	1 2 3 4	
Did you drift off into daydreams because you were sure you knew what the speaker was going to say?	1 2 3 4	
Did the speaker's mannerisms distract you?	1 2 3 4	
Were you distracted by what was going on around you?	1 2 3 4	

facilities are available you can pick up on facial and body language clues, but these are missing in a telephone interview. This skill improves with practice. Pilot interviews are vital, practise with a friend.

General

- ✔ Find a quiet room to call from, and get rid of all distractions.
- ✔ Lay your notes out around you and have two pens ready.

Research interviews

- ✔ Plan the call in advance. Organise your questions and comments so you can really concentrate on the responses and implications.
- ✔ Make notes of main points rather than every word, and leave time after the call to annotate and order the responses while it is fresh in your brain.
- ✔ Query anything you are unsure about. Be certain you understand the callers' nuances.
- ✔ Show you are listening and interested without interrupting, using 'yes', 'mmm', 'OK' and 'great'.
- ✔ Search for verbal clues, like a changed tone of voice, to 'hear between the lines'.
- ✔ Don't think you know all the answers already. If you disengage, the interviewee will become less engaged, less enthused and be a less productive informant.
- ✔ Curb your desire to jump in and fill pauses, let the speaker do most of the talking. Silences are OK.

Job interviews

- ✔ Prepare in advance as you would for a visit to a company, by researching the company background and position.
- ✔ Make notes of points as you speak, and query anything you don't understand.
- ✔ Be enthusiastic! All geographers have lots to offer.
- ✔ Be formal in your conversation, there is a job in prospect. It is easy to drop into a colloquial, conversational mode as if chatting to a friend, which you would not do in an office interview.

9.4 ONE-TO-ONE INTERVIEWS

Most of the tips for telephone interviews apply equally to personal interviews. Choose locations where you will not be interrupted. Take a coffee break in a long session to give both your own and your interviewee's brain a break. Remember people can speak at about 125 words a minute, but you can listen and process words at 375 to 500 words a minute so it is easy to find your brain ambling off in other directions. Don't wool gather!

Jumping to conclusions in discussion is dangerous, it leads you to switch off. Possibly the speaker is going in a new direction, diverting to give additional insights. Watch out for those 'yes but . . .' and 'except where . . .' statements.

Got a difficult customer? Let them talk, they will feel in charge and get the idea you agree with their discourse, which you may or may not.

Top Tips

- Really good listeners encourage the speaker by taking notes, nodding, smiling and looking interested. Do the Desmond Morris open, relaxed, friendly posture bit. Verbal feedback is often better as a statement that confirms what you have heard, rather than a question which will probably be answered by the speaker's next statement anyway. 'Did you mean . . . ?' or 'Am I right in thinking you are saying . . . ?' Unhelpful responses include yawning, looking out of the window, writing shopping lists and going to sleep. Relating similar personal experiences or offering solutions does not always help as, although you think you are offering empathy or sympathy, it may appear that you just turn any conversation around to yourself.
- When you are listening, interruptions need sensitive management. If you answer the phone or speak to the next person, the person you are speaking to will feel they are less important than the person who interrupts. If you do this to someone in business, they are very likely to take their business elsewhere. So turn the phone off and shut the door.
- If you are emotionally involved you tend to hear what you want to hear, not what is actually said. Remain objective and open-minded.
- Keep focused on what is said. Your mind does have the capacity to listen, think, write and ponder at the same time. There is time to summarise ideas and prepare questions, but it does take practice.
- Make a real attempt to understand what the other person is saying.
- Think about what is not being said. What are the implications? Do these gaps need exploration?

9.5 References and Further Reading

Bolton, R. 1979 *People Skills: how to assert yourself, listen to others and resolve conflicts*, Simon and Schuster Ltd., New York.

Brownell, J. 1986 *Building Active Listening Skills*, Prentice Hall, New Jersey.

Gushgari, S.K., Francis, P.A. and Saklou, J.H. 1997 Skills Critical to Long-term Profitability of Engineering Firms, *Journal of Management in Engineering*, **13**, 2, 46–56.

Lindsay, J.M. 1997 *Techniques in Human Geography*, Routledge, London.

What do you call someone illegally disposing of waste?

Jack the Tipper.

10 ORAL PRESENTATIONS

Tension's what you pay when someone's talking.

Most geography degrees are littered with speaking opportunities. This is a really good thing! Somewhere on your CV you can add a line like 'During my university career I have given 25 presentations to audiences ranging from 5–65, using OHTs, slides and inter-active computer displays.' This impresses employers, but many geography graduates fail to explain that they have had these opportunities to hone this very saleable skill. Speaking lets you get used to managing nerves, and makes you familiar with the question and answer session that follows. So grab all opportunities to practise your presentation skills. They all count, from 5-minute presentations in tutorials to seminars and mini-lectures.

You need a well-argued and supported geographical message to enrapture (maybe) the audience. Other chapters explain how to get the information together, this one is about practical presentation skills. The four most important tips are:

1. Suit the style and technical content of your talk to the skills and interests of the audience. Making the content accessible, so the audience wants to listen, will encourage a positive response.

2. Buzan and Buzan (1993) show that people are most likely to remember:
   (a) Items from the start of a learning activity.
   (b) Items from the end of a learning activity.
   (c) Items associated with things or patterns already learned.
   (d) Items that are emphasised or highlighted as unique or unusual.
   (e) Items of personal interest to the learner.

Tailor your presentation accordingly. Help the audience to think and understand.

3. Get the message straight in your head, organised and ready to flow. Lack of confidence in the content ➜ insecure speaker ➜ inattentive audience = Bad presentation.

4. Remember the audience only gets one chance to hear you, so pare down a short presentation to the essential points. Make clear links between points and add brief, but strong, supporting evidence.

You must have a plan, and stick to it on the day. Basically it is down to PBIGBEM (Put Brain In Gear Before Engaging Mouth).

10.1 Style

- **Can I read it?** Reading a script will bore both you and the audience. Really useful tutors will remove detailed notes and ask you to 'tell the story in your own words'. You are allowed bullet points on cards, to remind you of the main points but that is all. Illustrate your talk with maps, equations and diagrams on OHTs (overhead transparencies), but remember that reading OHTs aloud is another cop out.

- **Language** A formal presentation requires a formal speaking style. Try to minimise colloquial language, acronyms and paraphrasing, and limit verbal mannerisms like the excessive use of Hmm, Umm, err, and I mean. Put new words or acronyms on a handout or OHT.

- **Look into their eyes!** Look at the audience and smile at them. If they feel you are enthusiastic and involved with the material, they will be more involved and interested.

- **How fast?** Not too fast, slower than normal speech, because people taking notes need time to absorb your ideas, to get them into their brains and onto paper. The ideas are new to the audience. Watch the audience to check if you are going too fast, or they are falling asleep.

- **Stand or sit?** Position is often dictated by the room layout and normal practice. Given the choice, remember that sitting will encourage the audience to feel it is a less formal situation and one where it is easier to chip in and comment.

- **How loud?** So that the audience can hear, but do not feel you are shouting. Ask someone you trust to sit at the back and wave if you are too loud or too quiet, getting it right takes practice. Tape record yourself sometime. When you have finished laughing at the result, have a think about whether you speak at the same pace and pitch all the time. Changing pace and pitch, getting excited about the material and showing enthusiasm are all good techniques for keeping your audience attentive and involved.

- **What about repeating material?** You can emphasise primary points by repetition, or by simultaneously putting them on an OHT or board, but only repeat the important points.

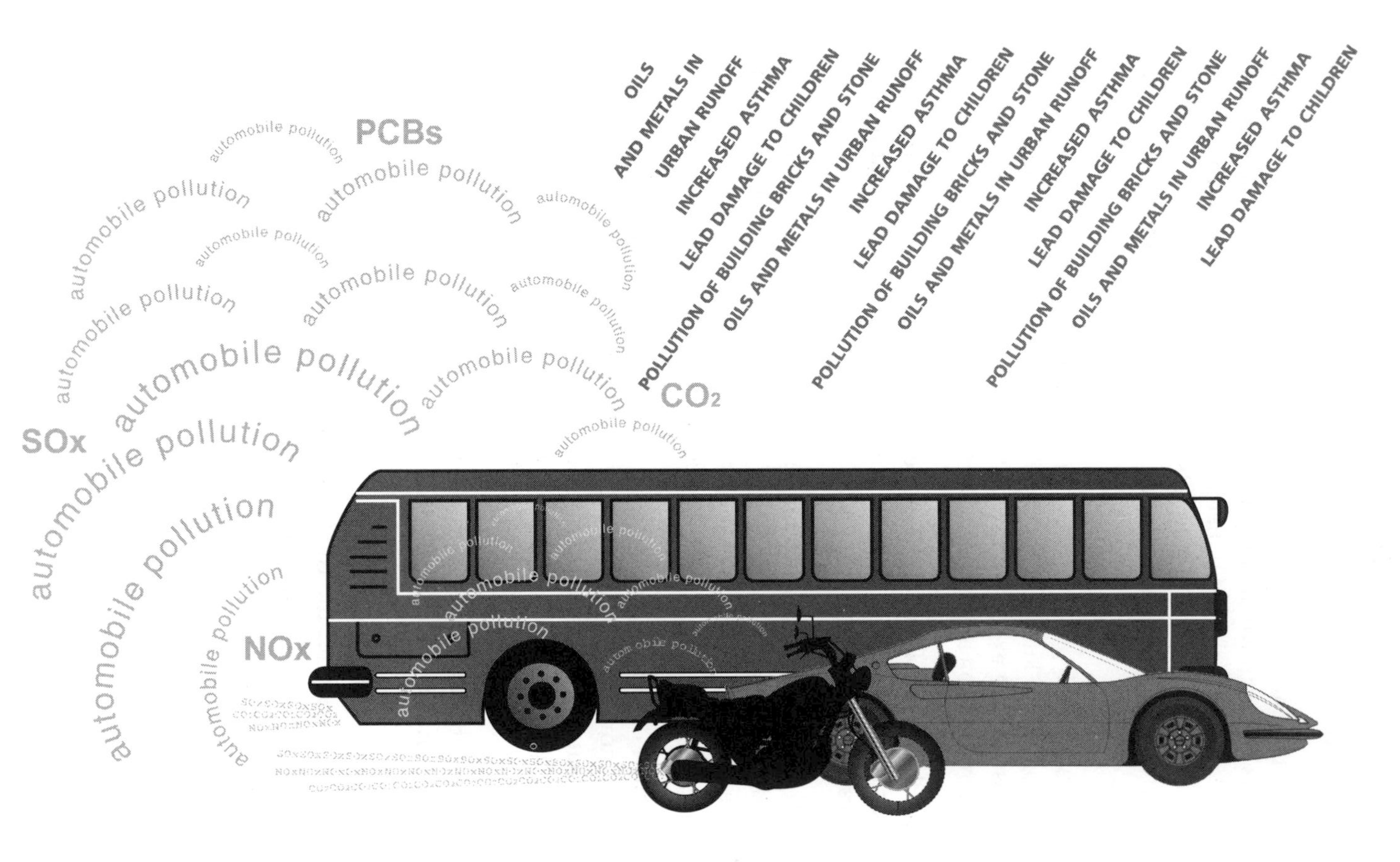

Figure 10:1 Presentation outline

10.2 Crutches (OHT, slides, flip charts)

Audiences need to understand your message. Visual assistance could include some or all of:

- ☺ A title slide that includes your name and (e-mail) address, so the audience knows who you are and where to find you!!
- ☺ A brief outline of the talk, bullet points are ideal, but adding pictures is fun, just make sure the message is clear. Is Figure 10.1 clear enough or OTT (over the top!)?
- ☺ A map, this is geography, tell the audience where you are talking about.
- ☺ Colours on complex diagrams help to disentangle the story.
- ☺ Graphs and pictures.
- ☺ Finally, a summary sheet. This may be the second, outline, slide shown again at the end, or a list of the key points you want the audience to remember.

OHT

An OHT (overhead transparency) is shown on an OHP (overhead projector). Preparing good OHTs and managing OHPs takes practice.

What does not work: Small writing, too many colours, untidy handwriting, and writing from edge to edge, leave a 5cm margin at least. Misuse of, or inconsistent Use of capitals Doesn't help Either.

Good Practice includes:

- OHTs prepared carefully in advance.
- SPELL CHECK EVERYTHING.
- Print in large font (18pt+) and then photocopy onto overhead film. (Putting overhead film through a computer printer will work with a few machines, usually you bust the printer.)
- A cartoon, clipart item or picture will make a message more memorable, and lighten the atmosphere if part of the material is seriously technical or a tad dull.
- Keep writing **VERY BIG** and messages short!
- It is fine to use diagrams and maps from books and papers, but usually you need to **enlarge** them and always cite their source.
- Some audience members will be colour blind, so avoid green and red together, black with blue, and black on red. Yellow and orange do not show up well in

large lecture theatres, orange and brown are not easy to distinguish from 30 rows back.

- Number your OHTs in case your 'friends' shuffle them or you drop them.
- If you have lots of information, and time is short, give the audience a handout with the detailed material. Use the talk to summarise the main points.

Before giving a talk, investigate the projector, find the on/off switches, the plug and how to focus the transparency. 'Sod's law of talk giving' says that the previous speaker will breeze in, move the OHP, give a brilliant talk and leave you to reset the stage while the audience watches you!! KEEP COOL, and DON'T PANIC. Have a pen or pencil handy to act as a pointer. Pointing at the transparency on an OHP shows greater cool than demonstrating your javelin technique with a pointer at a screen beyond your reach.

Slides

Make sure the room has a projector, can be blacked out and there is a carousel available. Run the slides through the projector in advance to ensure they are the right way round. Mixing overhead and slide projectors is a nightmare for novices and always wastes time in a talk. Try to use one medium or the other, so if you want to use slides get the title and talk outline slides onto film as well. Slides can be made from computer graphics package images. Just watch the cost and time for production. OHTs are fine for most purposes except very large audiences (over 300).

Flip charts

Again **Write Big** and make sure your pens are full of ink. Left- and right-handed persons need the chart in slightly different positions. Check beforehand what the audience can see, and adjust your position accordingly. If spelling is a problem, flip charts are a BAD IDEA. Irritatingly, the audience will remember your spelling error rather than your message.

10.3 What to avoid

☹ **Getting uptight.** All speakers are nervous. THIS IS NORMAL. Take deep breaths and relax before the event. If you are well prepared there is the time to walk slowly to the event and have a coffee beforehand. Get to the session early enough to find the loos, lights and seats. Ensure that the projectors, computer display system, flip chart, handouts, notes and a pen are in the right place for you and the audience. Then take a lot of deep breaths. YOU WILL BE FINE.

☹ **Showing the audience you are nervous.** Ask your mates to tell you what you do when speaking. Everyone gets uptight, but fiddling with hair, pockets, clothes, keys, pencil, ears and fingers distracts the audience. They watch the mannerisms and remember very little of the talk. Practise speaking.

☹ **Overrunning** As a rule of thumb, reckon you can speak at about 100–120 words per minute. Practise in advance with a stopwatch (there is one on most cookers), reducing the time by 10 per cent.

10.4 Handling sticky situations

☹ **Late arrivals.** They should be ignored, unless they apologise to you, in which case smile your thanks. Going back interrupts your flow and irritates the people who arrived on time.

☹ **Time is up and you have 20 more things to say.** This is bad planning and usually only happens when you have not practised the material aloud. You can read your material much, MUCH, faster than you can speak it. However, if you do run over time then either skip straight to your concluding phrases OR list the headings you have yet to cover and do a one-line conclusion.

☹ **Good question, no idea of the answer.** Say so. A phrase like 'That's a brilliant thought, not occurred to me at all, does anyone else know?' will cheer up the questioner and hopefully get others talking. If you need time to think, offer variations on 'I am glad this has come up …' , 'I wanted to follow up that idea but couldn't find anything in the library, can anyone help here?', 'Great question, no idea, how do we find out?',

☹ **The foot–mouth interface,** exemplified by such geographical classics as 'improved the engineering of buildings so they stood a better chance of collapsing in an earthquake.'; 'The gravity model can be heavy going to program'; 'What we found was that the desert was really, really dry, quite arid in fact.' They do happen.

Top Tips

- Self-assess your bathroom rehearsals against your department's criteria, or use the guidelines in Figure 10.2 to polish a performance.
- Use geographical language. Try not to use colloquial speech or to substitute less than technical terms, 'The results seemed a bit iffy.' 'There were loads of people …' or '… clouds, mmm, yes they are fluffy ones'. Strive in presentations to use a formal, technically rich style.
- Practise your talk on the parrot, the bathroom wall and … Speaking the words aloud will make you feel happier when going for the Gold Run. It also gives you a chance to pronounce unfamiliar words like equifinality, terretorialisation or *Rhytidiadelphus squarrosus*, and get them wrong in the

privacy of your own bus stop. If in doubt about pronunciation ask someone, the librarian, your mum, anyone.

- Most marks are given for content, so research the geography thoroughly.

Oral Presentation Criteria

Name .... Title of Talk ....

		Great	Middling	Oh Dear!	
Rendition					
Speed	Spot on				Too fast/slow
Audibility	Clear and Distinct				Indistinct
Holding Attention	Engaged audience				Sent them to sleep
Enthusiasm	Ebullient delivery				Boring delivery
Substance					
Organisation	Organised, logical structure				Scrappy and disorganised
Pertinence	On the topic				Random and off the point
Academic Accuracy	Factually spot on				Inaccurate
Support Materials					
Suitability	Tailored to talk				None, or off the topic
Production Quality	Clear				Unclear
Treatment	Professionally presented				Poorly integrated and presented
Question time					
Handling queries	Thoughtful answers				Limited ability to extend the discussion
Adaptability	Coped well				Limited, inflexible responses
Teamwork					
Co-ordination	Balanced, team response				Unequal, unbalanced response

Figure 10:2 Oral presentation assessment criteria

10.6 Further reading

Buzan, T. and Buzan, B. 1993 *The Mind Map Book*, BBC Books, London.
Sides, C.H. 1992 *How to Write and Present Technical Information*, (2nd edn), Cambridge University Press, Cambridge.
Young, C. 1998 Giving Oral Presentations, *Journal of Geography in Higher Education*, Directions, **22**, 2, 263–8.

11 DISCUSSION

Why is it that when you discuss Murphy's Law, something always goes wrong?

Geographers develop their research abilities through discussion in workshops, tutorials and seminars. Talking through the details of a topic leads to a greater degree of understanding and learning. In most jobs, being able to discuss topics calmly, fairly and professionally is essential, so discussions are valuable opportunities to practise.

To learn effectively from discussions there needs to be a relaxed atmosphere where people can think about the content, and note what others are saying. Ineffective discussions occur when people are worrying about what to say next, and running through it mentally, rather than listening to the person speaking. Some of this is nerves, which will calm down with practise, but in the meantime preparing fully is the best way to lower your stress levels. You will have background and specific information to share.

Be positive about seeking the views of others and value their contributions. Employ open-ended questions, those which encourage an elaborated, rather than a brief yes or no answer. 'What are the main ... resort facilities of Penang?', is an open-ended question and more useful than 'Do you like ... Penang?' 'What do you think about ... green issues/global warming/transport in New Delhi?' are good questions, but 'What do you think ...' is a bit general. Use phrases like 'What are the advantages of ... global warming?' or 'Do you consider ... the Internet's global, unrestricted basis will influence the actions and attitudes of individuals?' or 'What do you value about ... the lake and its environs as a recreation resource for local people?'

Keep up the quality of argument in discussion. For example, you might discuss the reliability of visual evidence as a geographical source. You might make a general point like 'People have tended to believe the pictorial record presented on film, or in photographs, believing the general tenet that the camera cannot lie.' This is a general argument that would be strengthened, getting more marks, by adding references and examples as you speak. You might say, 'However Crang (1997), suggests that the photographer and film maker selects shots and views to convey particular messages and therefore the message may be biased by his choices.' Adding an example would increase memorability. For top marks, take examples from more than one source.

11.1 Types of discussion

Brainstorming

Brainstorming is a great way of collecting a range of ideas and opinions and getting a group talking. The process involves everyone calling out points and ideas. Someone keeps a list, maybe on a flipchart so everyone can see. A typical list has no organisation, there is overlap, repetition and a mix of facts and opinions. The art of brainstorming is to assemble ideas, including the wild and wacky, so that many avenues are explored. The points are reordered and arguments developed through discussion, so that by the end of a session they have been pooled, ordered and critically discussed. Have a go at **Try This 11.1** as an example of working with a brainstormed list , and use **Try This 11.2** with your next essay.

TRY THIS 11.1 – Working with a brainstormed list

Brainstorming produces a list of ideas with minimal detail and no evaluation. This list was compiled in five minutes during a coastal fieldtrip. Items overlap and are repeated, facts and opinions are mixed, and there is no order. Take five minutes to categorise the items. A possible sort is given on p.259.

Human impacts on the coastal landscape at ....:

Concreting the slopes to prevent soil erosion.	Sea wall prevents undercutting.
Rocks at cliff foot dissipate wave energy.	Groynes prevent longshore drift.
Fine and coarse netting, geotextiles, used on clay slope to help stabilise the soil.	
Bolts into the rock face to increase stability.	Tourist facilities destroy foreshore view.
Rock armour installed at the foot of recent slip.	Movement of handrails.
Vegetation eroded from wetter clay slopes.	Tennis courts on levelled slip.
Evidence that paths across the slips are regularly re-laid, cracked tarmac, fences reset.	
Café for tourists and visitors at foot of slip area.	Soil overlying clay rock.
Wall to prevent boulders from cliff the road.	Tree planting on recent and older slip hitting sites.
Drainage holes in retaining walls to reduce soil water pressure.	
Vegetation worn on paths and alongside long views.	Car parking demands, unsightly in paths.
Promenade held up by an assortment of stilts, walls and cantilever structures where slope below has been undercut.	Litter needs collecting from beach.
Road built over old stream, presumed piped underground.	
Castle fortification, some walls now undermined and eroded.	
Hotel, amusement arcade and tourist shops – noisy and unsightly.	
Harbour has mix of marina, fishing and tourist fairground rides.	
Path covered by slumping clay.	Repairs to sea wall.
Victorian swimming pool, in need of repair.	Car pollution.
Car parking unscenic.	Caravan park on cliff.
Varied quality of coastal path, not always well sign posted.	

TRY THIS 11.2 – Brainstorm an essay plan

Using your next essay title or a revision essay, brainstorm a list of ideas, alone or with friends, including references and authors, and use this as the basis for essay planning. Brainstorming 'what I know already' at the start of essay planning, can indicate where further research is required.

Role play exercises

These may involve the simulation of a meeting, as for example where environmental conflicts are explored through a planning enquiry forum. You will prepare a role in advance, not necessarily a role you would agree with personally. Procedure depends on the type of topic. It may lead to a 'public enquiry' decision given by the Chairman, or a 'jury' vote.

Debate

The normal format for a debate presupposes that there is a clear issue on which there are polarised opinions. A motion is 'put' forward for discussion. It is traditionally put in the form 'This house believes that . . .'. One side proposes the motion, and the other side opposes it. The proposer gives a speech in favour, followed by the opposer speaking against the motion. These speeches are 'seconded' by two further speeches for and against, although for reasons of time these may be dispensed with, in one-hour debates. The motion is then thrown open so everyone can contribute. The proposer and the opposer make closing speeches in which they can answer points made during the debate, followed by a vote. Geographical issues are often not clear-cut and a vote may be inappropriate, but a formal debate is a useful way of exploring positions and opinions, and of eliciting reasoned responses.

Oppositional discussion

Oppositional discussion is a less formal version of debate, in which each side tries to persuade an audience that a particular case is right and the other is wrong. You may work in a small group, assembling information from one point of view, and then argue your case with another group which has tackled the same topic from another angle. Remember that your argument needs supporting with evidence, so keep case examples handy.

Consensual discussion

Consensual discussion involves a group of people with a common purpose, pooling their resources to reach an agreement. Demonstrations of good, co-operative, discussion skills are rare, most of the models of discussion on TV, radio, and in the press are set up as oppositional rather than consensual. Generally you achieve more through discussing topics in a co-operative spirit and one of the abilities most sought after by employers of graduates is the ability to solve problems through teamwork.

Negotiation

Negotiation, coming to an agreement by mutual consent, is another useful business skill. One practises and improves negotiation skills in informal everyday activities like persuading a tutor to extend an essay deadline, getting a landlord to do repairs or persuading someone else to clean the kitchen. In formal negotiations do the following:

- ✔ Prepare by considering the issues in their widest context, in advance.
- ✔ Enumerate the strengths and weaknesses of your position. It reduces the chances of being caught out!
- ✔ Get all the options and alternatives outlined at the start. There are different routes to any solution and everyone needs to understand the choices available.
- ✔ Check that everyone agrees that no major issue is being overlooked, and that all the information is available to everyone.
- ✔ Appreciate that there will be more than one point of view, and let everyone have their say.
- ✔ Stick to the issues that are raised and avoid personality-based discussion, that person may be an idiot BUT saying so will not promote agreement.
- ✔ Assuming decision deadlines are flexible, break for coffee, or agree to meet again later, if discussion gets over-heated.
- ✔ At the end, ensure everyone understands what has been decided by circulating a summary note.

There are many books on discussion, assertiveness, and negotiation skills, see Drew and Bingham (1997), or do a keyword search to see what your library offers.

Top Tips

- Being asked to start a discussion is not like being asked to represent your country at football. You are simply 'kicking off'. Make your points clearly and 'pass the ball' promptly. Focus thoughts by putting the main points on a handout or OHT.
- Don't wait for a 'big moment' before contributing. Ask questions to get a topic going.
- Don't be anxious about the quality of your contributions. Get stuck in. Early in a discussion everyone is nervous and too concerned about his or her own contribution to be critical of others.

- Keep discussion points short and simple.
- Use examples to illustrate and strengthen your argument.
- Share the responsibility for keeping the group going.
- Have a short discussion before a tutorial to kick ideas about. Meet in the bar, over coffee or supper, somewhere informal.

11.2 Electronic discussions

It is not always possible to get people together in the same room for a discussion. Electronic discussions can solve groupwork timetabling problems, and are especially useful for part-time and students off-campus on years abroad or in work placements. They are also good practice for business discussions. One advantage of electronic discussions is that you can build research activities into the process. Having started a discussion, you may realise that you need additional information. You can find it and feed it in as the discussion progresses. Electronic discussions are held via e-mail and bulletin boards, or you may get involved in computer conferencing. The methodology depends on the local technology (Aldred 1996; Rapaport 1991).
Here are some suggested ground rules to make electronic discussions successful, with possible times and numbers for an e-mail tutorial discussion in brackets.

- Agree a date to finish (2 weeks).
- Everyone must make a minimum number of contributions (three).
- Agree to read contributions every x days (two or three).
- Appoint someone to keep and collate all messages so there is a final record.
- Appoint a 'devil's advocate' or 'pot stirrer' to ask awkward questions and chivvy activity.
- Ask someone to summarise and circulate an overview at the end (the basis of a group report for everyone to amend if the discussion is assessed).
- Be polite. In a conversation you can see and hear when someone is making a joke or ironic comment. The effect is not always the same on paper.
- Where further research is required, attempt to share tasks evenly.
- Replying instantly is generally a good idea, that is what happens in face-to-face discussion, and first thoughts are often best.

Some people 'lurk' quietly, listening rather than commenting, which happens in all discussions. Point two, above, should overcome this issue to a certain extent. One of the more off-putting things that can happen in an electronic discussion is someone writing a 3000-word essay and mailing it to the group. This is the equivalent of one person talking continuously for an hour. It puts off the rest of the group, they will feel there is little to add. Try to keep contributions short in the first stages. One good way to start is to ask everyone to brainstorm 4–6 points to one person by the end of Day 2. These are collated, ordered and mailed around the group as the starting point for discussion and research. (See **Try This 13.1 E-mail in action** p 141).

11.3 Group management

The quality of discussion depends above all on the dynamics of each particular group. Some work spontaneously without any problems, others are very sticky. There are no hard and fast rules about behaviour in group discussions but here are some general points to consider. Meetings flow well when members do the following:

- ☺ Chair the discussion to keep it to the point, sum up, shut up people who talk too much, and bring in people who talk too little.
- ☺ Keep track of the proceedings.
- ☺ Inject new ideas.
- ☺ Are critical of ideas.
- ☺ Play devil's advocate.
- ☺ Calm tempers.
- ☺ Add humour.

Formal meetings have designated individuals for the first two tasks (Chair and Secretary), but in informal meetings *anyone* can take these roles at any time. Everyone is better at one or two particular roles in a discussion, think about the sorts of roles you play and also about developing other roles using **Try This 11.3**.

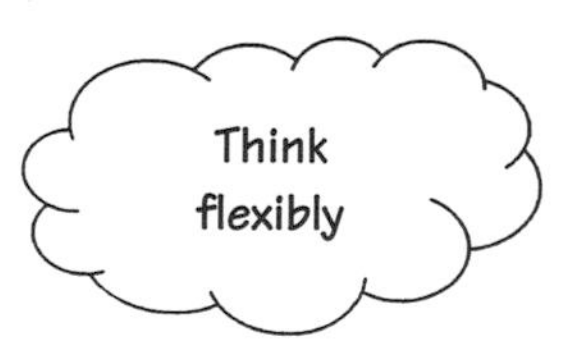

TRY THIS 11.3 – Discussant's role

Here is a list of the actions or roles people take in discussions (adapted from Rabow *et al.* 1994). Sort them into those which are positive and promote discussion, and those which are negative. (Suggested answers on p.259). How might you handle different approaches?

<table>
<tr><td>Offers factual information</td><td>Gives factual information</td></tr>
<tr><td>Speaks aggressively</td><td>Asks for examples</td></tr>
<tr><td>Encourages others to speak</td><td>Asks for reactions</td></tr>
<tr><td>Asks for examples</td><td>Seeks the sympathy vote</td></tr>
<tr><td>Is very competitive</td><td>Offers opinions</td></tr>
<tr><td>Helps to summarise the discussion</td><td>Is very defensive</td></tr>
<tr><td>Keeps quiet</td><td>Summarises and moves discussion to next point</td></tr>
<tr><td>Ignores a member's contribution</td><td>Gives examples</td></tr>
<tr><td>Mucks about</td><td>Diverts the discussion to other topics</td></tr>
<tr><td>Asks for opinions</td><td rowspan="2">Keeps arguing for the same idea, although the discussion has moved on</td></tr>
<tr><td>Is very (aggressively) confrontational</td></tr>
</table>

11.4 Assessing discussions

At the end of each term or as part of a Learning Log (see Chapter 2) you may be asked to reflect on your contribution to discussion sessions, and in some cases to negotiate a mark for it with your tutor. The attributes an assessor might check for are included in **Try This 11.4**. Assess friends, seminar and TV discussants on this basis, and consider what you can learn from those with high scores.

TRY THIS 11.4 – Assessment of discussion skills

Evaluate a discussant's performance on a 1–5 scale and note what they do well.

Assessment of Discussion Skills	1 (useless), 3 (average), 5 (brilliant) Comment on good points
Talks in full sentences	
Asks clear, relevant questions	
Describes an event clearly	
Listens and responds to conversation	
Discusses and debates constructively	
Speaks clearly and with expression	
Selects relevant information from listening	
Responds to instructions	
Contributes usefully to discussion	
Reports events in sequence and detail	
Is able to see both sides of the question	
Finds alternative ways of saying the same thing	
Listens to others, and appreciates their input, efforts and needs	

Reflect on the tips you can pick up from 4–5 point performers.

Having analysed what makes a good discussant, have a look at **Try This 11.5**.

Discuss Everything

TRY THIS 11.5 – Self-assessment of discussion skills

Level 1 students at Leeds brainstormed the following list of 'skills they needed to argue effectively'. Look down the list and select three items where you would like to be more effective. Do you have items to add to the list?

Now plot a strategy to work on each of these three issues at your next group discussion. (e.g. I will not butt in; I will ask at least one question; I will say something and then shut up until at least three other people have spoken).

Being open-minded	Staying cool
Listening to both sides of the argument	Being tolerant
Using opponents words against them	Using good evidence
Playing the devils advocate	Being willing to let others speak
Summing up every so often	Only one person talking at a time
Thinking before speaking	Being firm

Some final points

Getting better at discussion and argument needs practise, and hearing one's own voice improves one's self-confidence. You can practise in private. Listen to a question on a TV or radio discussion programme. Then turn the sound down, take a deep breath to calm down, and use it as thinking time. What is the first point? Now say it out loud. Subject matter is not important, get in there and have a go. Respond with two points and then a question or an observation that throws the topic back to the group or audience. That is a good technique because you share the discussion with the rest of the audience, who can contribute their range of views. You might want to tape a programme and compare your answer with the panellists, remembering to look at the style of the answers rather than their technical content.

Where points of view or judgements are needed, you may want to seek the opinions of people in different academic, social and cultural backgrounds and experience. Their views may be radically different from your own. Seminars, workshops and tutorial discussions in geography are explicitly designed to allow you to share these kinds of complementary views. To get the most out of a discussion or conversation, do the following:

- ✔ Be positive.
- ✔ Ask yourself questions, like 'How will this help me understand ... passenger transport pricing?'
- ✔ Make eye contact with the group.
- ✔ Give the speaker feedback and support.

- ✔ Aim to be accurate and on the point.
- ✔ Include geographical examples and references as you speak.

11.5 References and further reading

Aldred, B.K. 1996 *Desktop Conferencing: a complete guide to its applications and technology*, McGraw-Hill, London.

Crang, M. 1997 Picturing Practices: research through the tourist gaze, *Progress in Human Geography*, **21**, 3, 359–73.

Drew, S. and Bingham, R. 1997 Negotiating and Assertiveness, in Drew, S. and Bingham, R. (eds) *The Student Skills Guide*, Gower, Aldershot, 121–34 and 255–62.

Rabow, J., Charness, M.A., Kipperman, J. and Radcliffe-Vasile, S. 1994 *William Fawcett Hill's Learning Through Discussion*, (3rd edn) Sage Publications Inc., California.

Rapaport, M. 1991 *Computer Mediated Communications: bulletin boards, computer conferencing, electronic mail, and information retrieval*, John Wiley, Chichester.

Geo-cryptic crossword 1

Answers on p.259

Across

1 Mariners, just a loud one fixes prices (7)
5 Worth going to lawyer's board? (7)
9 Fruitfulness found in iron and German city (9)
10 Decoy put together to attract wild rodent (5)
11 Rough or fair he knows the vagaries are still magic too (13)
13 Cold pasture for glacier, no Dome here! (8)
15 Direct in to kingly play (6)
17 Early film, we hear, from Torquay? (6)
19 Rosa paid her way on migration (8)
22 Left a flexible impression (6, 7)
25 Arterial trunk route (5)
26 Turns sour if I identify a maple.(9)
27 Weakened detective played old guitar for 500 (7)
28 Might dig out the stinking camomile (7)

Down

1 Porous rock? Sounds like a rough break for the geologist (4)
2 Break down, again and again (7)
3 Mounds are broken for soldiers amusement (5)
4 The French précis gave vent through breathing hole (8)
5 'A blight on all your houses?' (3, 3)
6 The nicest cot reserved for those involved with structural geology (9)
7 Said bye to staying by the sea (7)
8 Find Quito a real location (10)
12 Up river from Tehran, Linda lost her end (10)
14 Anti-rent? so I become homeless (9)
16 the other way, it is a metal (8)
18 A generous politician (7)
20 Venerable tempo (3, 4)
21 No man is one? (6)
23 Having an intimate interest perhaps, in council (5)
24 German obliterated a network (4)

12 EFFECTIVE ESSAY SKILLS

I'm getting on with the essay, I started the page numbers yesterday morning, and they look really good.

Many of the chapters in this book start with some reasons why practising a particular skill might be a moderately good idea, and an analogy involving practising playing musical instruments or sport. Now demonstrate your advanced creative skills by designing your own opening. Create two well-argued sentences using the following words or phrases: reports, communication, language, persuasion, cheerful examiners, life-long, clarity; and reorganise to make a coherent sentence: with needs solos . practise, bagpipe As writing

The paragraph above may give you an insight into how annoying examiners find disjointed, half-written paragraphs, with odd words and phrases rather than a structured argument. Such paragraphs have no place in essays. If this is the first chapter you have turned to, keep reading!!

This chapter picks up some important points for geography essays, but many texts discuss writing skills in more detail. Mature students who have 'not written an essay in years' or those who did science A levels and 'haven't written an essay since ...' should find Russell (1993) or Fitzgerald (1994) valuable. Many universities run essay skills classes, and the fact that so much geography assessment is still essay-based should encourage *everyone* to go along. For tips on using English, grammar and spelling, Kirkman (1993) is particularly good, delightfully short and suitable for both human and physical geographers.

Tutors are often asked what a good answer looks like. The most helpful response involves comparing and evaluating different pieces of writing. This chapter includes some examples of student writing at different standards. You are asked to compare the extracts to get a feel for the type of writing you are expected to produce. Space restricts the selection, but you could continue by comparing essays with friends and other members of your tutor group.

All essays need good starts and ends, lots of support material and a balance of personal research and lecture based evidence. That usually requires an initial plan, some rethinking, writing, further research, and re-writing. Then a heavy editing session where the initial long sentences are cut down to shorter ones and paragraphs are broken up, so that each paragraph makes a separate point. The first version of anything you write is a draft, a rough and ready first attempt, requiring development and polish before it is a quality product. Most marks disappear because the first draft is submitted as the final product.

12.1 What kinds of essays are there?

The 'what do you know about ...' style essay should be disappearing from your life. University questions usually require you to *think* about information that you have researched and to *weave* it into an argument. You are asked to analyse, criticise, examine, and debate ideas in a structured way, using apt examples to illustrate your arguments. Essays that get high marks interweave lecture material with personal research findings and ideas. Facts from lectures, by themselves, are not enough. Painful but true. Reproducing the facts and arguments as presented in a lecture may get you a mark of 30–50 per cent. To get 50 per cent plus you need to show an examiner that you have thought about the issues. This involves adding other information gleaned through reading, sorting out what it all means for you, and re-stating the argument in your own words (*see* Figure 12.1). OK, that is my opinion. Ask your geography tutors what they think about this, get your own department's view.

You will increasingly be asked for *discussion* rather than *descriptive* essays. Compare these questions:

1. *Descriptive.* Outline and justify the methods of store location you would recommend to a retailer.
2. *Discussion.* Store location is a retailer's minefield. Discuss.
3. *Descriptive.* Describe the impact of socialist policies on China's cities.
4. *Discussion.* The contradictions in the market socialist policies of China are written across the face of its cities. Discuss.

The descriptive essay title may have pointers to the structure, and type of answer required. The discussion essay needs more thought and planning; you must establish your own structure, and write an introduction to signpost it to the reader. This should be followed by linked arguments supported by evidence, leading to a conclusion that follows from the points you have made. Including material that veers off at a tangent, or is irrelevant, or presenting evidence in ways that do not really support your case, loses marks.

What are the supporting ideas?

Many geography essays involve a question with no right answer. You consider the various dilemmas and decisions that are the essence of real situations, provide evidence and reach a balanced conclusion. Questions like 'Compare the strategies homeless men and women develop to survive in urban areas', 'Lay people should not have a role in planning decisions. Discuss', 'Everyday life is worthy of serious geographical research attention. Discuss', 'Consider whether global climate modelling presents an impossible challenge' and 'There should be no more nuclear power stations'.

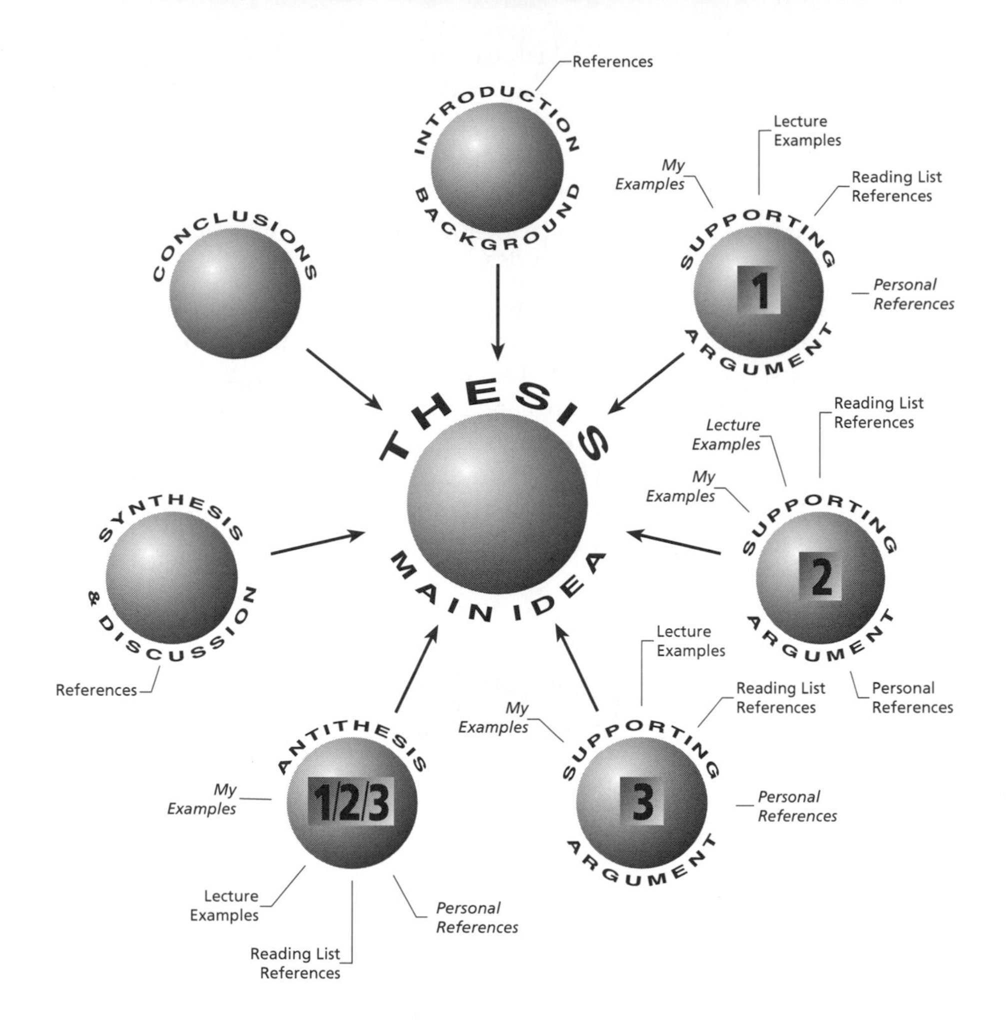

Figure 12:1 Structuring an essay

12.2 Carefully analyse essay titles

The wording of a question can give you guidelines for your answer. Take time to analyse the question. For example:

Discuss the causes, effects and methods to reduce hazards associated with earthquakes.

What are the keywords? Anyone who thinks that this is an essay about earthquakes is in serious trouble. It is an essay about fire, flooding, landslides, tsunami, the collapse of bridges and disruption to communications, a hazards essay. Earthquakes get a passing reference in exemplification, as for example 'The collapse of housing stock in the Turkish earthquakes of ...' or 'The severing of network links where roads collapsed during the San Francisco earthquake led to ...', they are not the essay's focus. Equal weight needs to be given to the three main sections of the essay – 'causes', 'effects' and 'methods to reduce', with each aspect discussed in relation to each hazard.

Describe and critically evaluate the factors that are considered to drive glacial cycles.

What are the keywords? The problem with this essay is that there is too much information. The influence of plate tectonics, orogeny, the raising of the Tibetan plateau and consequent feedbacks to climate, changing global CO_2 and temperature balances, sea temperatures and currents, volcanic dust, astronomical theories including Milankovitch, equinoctial procession, orbital eccentricity, sunspot and solar flare theories. There is an excess of Knowledge material. Most students attempt to include all the knowledge and thereby throw away the 50 per cent of the marks for the 'Critically Evaluate' section.

Now look at **Try This 12.1**.

- Assume all essay questions include the phrase **'with reference to specific geographical examples'**.

TRY THIS 12.1 – Keywords in essay questions

What are the keywords and potential pitfalls with the following essay titles? Answers on p.260.

1. Communications, internal commerce and energy are the sectors that are usually identified as the most serious 'bottlenecks' in contemporary Chinese development. Explain the weaknesses in one of these sectors.
2. To what extent are growth and change impeded by archaic social structures in either Latin America or a selected country in Latin America?
3. Critically explain how Andrews (1983) derived his bedload entrainment function.
4. Account for the rising profile of the service industry and assess the limits to the internationalisation of services.
5. Evaluate the role of mass media reportage of environmental issues.

12.3 Effective introductions

A good introduction serves two purposes. It outlines the general background or position and signposts the structure and arguments that follow. It gives the reader confidence that you are in command of the topic. There is no reason for the introduction to be long, 100–200 words or half a side is usually plenty. If the introduction is longer, consider editing it and putting some material later in the essay.

Introductions require some thought. *Planning time is vital.* A provocative, headline opening, grasping the imagination of the reader is a good wheeze, BUT it must be integral to the essay. Lengthy case material and examples are not usually good opening material. If there is an example in the first paragraph, limit it to one sentence, further details belong in the body of the essay. Here is a reasonably good introduction (it could be improved):

Explain with examples, why river flows in urban channels are rarely natural.

Urbanisation and the consequent paving of natural surfaces with cement, roofs and tarmac have changed the natural water balance within cities, there is less infiltration, less soil storage and increased rapid surface runoff (Wellis and Mack 2010). Urban flooding can be both personally inconvenient and economically disruptive, so runoff is channelled efficiently through drains and sewers to minimise surface inundation. The demand for water in the city is high, and in developed countries particularly, this demand has increased through the twentieth century as lifestyles have changed (Gluggit 2010). Water from local sources is used where possible, and transferred from adjacent catchments to augment supply as required. Effluent is normally treated locally and disposed to the nearest watercourse. These competing demands for extraction and disposal of water, combined with maintaining minimum flows in drought periods and controlling flood flows mean that urban channel flow is managed and regulated rather than natural. These demands are not always compatible; a strategic planning issue for water managers. The issues and conflicts that arise will be discussed in the context of a range of case studies from the UK and USA.

This introduction summarises some of the major interlocking issues, indicates that there are no simple answers and that examples will be taken from two developed countries.

This next example contains a series of true statements, but the style of the essay is not indicated. AVOID THIS approach please.

Discuss the economic impact of floods in urban areas.

Over history many very important urban developments have grown up besides or around fluvial channels which are susceptible to flooding. Over this same period the inhabitants of those urban areas have tried to manipulate the natural channel so as to reduce the risk of flooding and thus economic impact on that area. Although flooding may take place at any point along the channel's path, it is in urban areas that the greatest density of housing is affected, thus maximum disruption takes place and the largest economic impact is felt.

The first sentence is long and woolly, it could read 'Historically, flood prone river side sites have been chosen for settlement'; 11 words, not 20. Similarly, the start of sentence 3 could read 'Although flooding may occur anywhere along the channel';

8 words rather than 12. There is no indication in this paragraph of the topics the essay will address, no mention of direct and indirect costs, insurance and liability, planning legislation to control floodplain development, modelling insights from hydrological and economic aspects and case examples.

A more headline style start might read:

Who pays if your business is flooded by storms, snowmelt, hurricanes, water backed up from overloaded storm sewers, or the failure of a dam or embankment? More than 90 per cent of the disastrous floods in the USA are not declared as federal emergencies. They do not qualify for financial assistance. If your business is flooded, the costs are your problem. ...

Or

You do not have to live on a floodplain to be flooded in an urban area. Flooding following unusually large storms, the backing up of sewers, rapid snowmelt, collapse of floodwalls, dams or embankments, or the failure of a mains water supply pipeline can hit any home or business.

Pretend to be an assessor and take a critical look at **Try This 12.2**. Then try the same technique with your recent essays.

TRY THIS 12.2 – Evaluate an introduction

Pretend to be an examiner looking at these four introductory paragraphs to the same essay. You need no knowledge of Canada or Wetlands; a small gap in your knowledge will let you concentrate on the main issues for good opening paragraphs. Questions to ask include:

- Is the general case outlined and explained?
- Does the introduction indicate how the author will tackle the essay?
- Is the language suitably technical (or 'grown up' as a tutee once described it)?

The essay title is horribly general, which makes a good opening particularly important. Some comments are on p.260.

Discuss the impact of man on Canadian wetland landscapes in the last 100 years.

Version 1

Wetlands can be described as transition zones between terrestrial landforms and water bodies. They resemble both uplands, due to their ability to support emergent plants, and aquatic regions because of the domination of the areas by water. Even though these highly productive areas contribute 14 per cent of Canada's total land surface, they were not thought of as having potential until the end of the last century because they made transport difficult whilst at the same time harbouring many deadly diseases.

Version 2

Wetlands perform a range of valuable functions including reducing flooding, filtering polluted waters, defending shorelines and as habitats for many species including migrant birds, which makes them landscapes worth retaining. They are found wherever the watertable is at or near the surface for most the year and include salt and freshwater marshes, bogs, sloughs, fens and the margins of lakes and rivers. They are dependent for their structure and function on a sustained hydrological and nutrient status, but this also makes them vulnerable. A change in the water supply through drainage, drought or flooding will upset the ecology, as will a sudden increase or decrease in nutrient supply (Verboggi 2010). Recognition of the value of wetlands is a recent phenomenon. In the past they were considered as wastelands, dangerous quagmires that could trap animals and the habitat for insects that cause diseases such as malaria. Early settlers in Canada could afford to ignore wetlands, there were more promising lands to farm, but as the pressure on land grew through the twentieth century wetlands were increasingly reclaimed. In the past 100 years a diversity of competing interests has led to the reclamation for agriculture, drainage schemes, urban expansion, forestry, and peat extraction for horticulture and fuel (Albertus and Brit-Columb 2010). The competing issues and impacts are illustrated here through a range of case examples from Arctic to continental wetlands highlighting the vulnerability of the sites and the problems of protection and reclamation. However, despite the loss of 70 per cent of Canada's wetlands there are good examples of conservation and site protection exemplified through RAMSAR and other sites.

Version 3

Due to the ever-increasing population and rapidity of land use change the prevalence of wetland areas (wetlands are a transition between landforms and water bodies) is depleting. Within urban areas wetlands are often the last to be developed and as cities have expanded through this century there has been a rush to fill these in too. It is man's impact that particularly over the last century has influenced them most. Many have been depleted due to their site near to main rivers, coasts and bays, and therefore have been developed due to their necessary use for transportation. The conversion of lakes and reservoirs has also added to this. This essay will examine some of the impacts on wetlands in the last 100 years.

Version 4

Canadian wetlands are extensive but vulnerable to exploitation and reclamation (Environment Canada 2010). The pressures of agriculture, drainage schemes, urban development, forestry and peat harvesting have significantly reduced their extent and this has had an impact on the wildlife which use wetlands as permanent or migration habitats. This essay will assess the extent and processes of change using examples from wetlands across Canada, and then focus on the particular issues raised by the Copetown Bog near Ontario. As this case reveals, there are grounds for optimism as the significance of wetlands, and the need to conserve them, are nationally and internationally recognised.

12.4 The middle bit!

This is where to put the knowledge and commentary, developed in a logical order, like a story. *You need a plan.* Sub-headings will make the plan clear to you and the reader, but whether you use sub-headings mentally, or mentally and physically is a matter of personal taste. It may be worth asking your tutors if they mind. Most tutors will not, but some are vitriolically against sub-headings in essays. Find out first.

Examples are the vital evidence that support the argument. Where you have a general question like 'Describe the impact of satellite TV on global marketing,' aim to include a range of case examples, BUT if the question is specific, 'Comment on the impact of Sainsbury's marketing strategy' then the essay needs to mostly be about Sainsbury's. In the 'global marketing of TV' essay, remember to make the examples global. There is a temptation to answer questions like this with evidence from the USA, Europe and Japan, forgetting the developing world. My advice is to use as many relevant examples as possible with appropriate citations. Generally, many examples described briefly get more marks than one example retold in great depth. Examples mentioned in lectures will be used by 80 per cent of your colleagues. Exhilarate an examiner with a new example.

Graphs, figures and pictures

Graphs, figures, and pictures should illustrate and support the arguments, and do so with fewer words than a description of the same material. Put them in. Label them fully and accurately, and refer to them in the text. If there is no instruction to a reader to look at Figure X, they will get to the end of your document without looking at it, and the inclusion is wasted. One advantage of labelling figures is that they can be cross-referenced from elsewhere, especially discussion and conclusion sections, in the essay.

Reference appropriately

All sources and quotations must be fully referenced (see Chapter 14 for details) including the sources of diagrams and data. Avoid the obvious: 'Water flows downhill (Kirkby 2010)' *OK* it does, and Kirkby may have said it as part of a rivers or erosion lecture but this is not really an appropriate example of referencing, whereas 'The incorporation of the area drained per unit contour length concept, a/s tanβ (Kirkby 1978), into the model allows ...' is a proper and highly desirable example of referencing.

12.5 Synthesis and Conclusion

A conclusion is the place you have reached when you are tired of thinking.

Synthesis

The discussion or synthesis section allows you to demonstrate your skill in drawing together the threads of the essay. You might express the main points in single sentences with supporting references, but this is just one suggestion. Here are two examples of synthesis paragraphs. In answer to the essay question: 'Evaluate the principal theories of infiltration', paragraphs 2–8 outlined the main theories with the equations, the knowledge section of the answer. Followed by:

Current understanding of the processes of infiltration is influenced by the three main theories outlined above. For dry and sandy soils with high permeability, the Green and Ampt (1921) model is appropriate. The Phillips equations (1937) have been widely and successfully used, for example by Engman and Rogowski (1976) in an overland flow routing model. In saturated and semi-flooded soils, Hillel's (1986) model provides a more accurate set of results than the Phillips approach. This is because permanent or semi-permanent ponding on the surface reduces the air infiltration element from the equation. Neither of these models is appropriate in permafrost areas and with frozen soils. For these environments the results of Dunne (1964), although limited to soils in Nebraska, seem to provide a more accurate model. In addition to these field approaches, the data from laboratory experiments by Jackson (1967) and Burt (1982) have improved our process understanding in detail. However, these approaches, although apparently successful, have yet to be tested at the field scale. At present there are a suite of approaches available to the hydrologist. However, s/he needs to select the technique that is most compatible with prevailing environmental conditions.

So what do you think? Lots of information, inter-woven references, useful information – this should get high marks and is a style to aim for. The last three sentences are very good. If the facts were right, it would get high marks. BUT this particular paragraph will be lucky to score 40/100. The authors are all hydrologists and some did the work referred to, some did not and all the years are wrong. What is worse, is that the most recent reference is 1986, has nothing happened since then? This is an unacceptably dated answer for 1999+. Examiners notice spoof and imaginary references.

On a cultural topic this example, with correct references, weaves in some definitions:

The old cultural geography of the Berkeley School, under Carl Sauer (1925), has been criticised for the superorganic view of cultures it held. The approach was morphometric, reading culture unproblematically from the landscape. Old cultural geography had little to say about the individual agents who make up a culture and it was because of this neglect that the approach was derided as essentially an American white male's vision, which ignored cultural specificity and nuance.

Culture has been described as one of the most 'difficult' words in the English language (Williams 1976). This difficulty is reflected in problems of defining when the 'new' cultural geography began, although Cosgrove (1992) suggests 1978. In contrast to the old cultural geography, Cosgrove considers that new cultural geography looks at the contemporary as well as the historical, rural as well as urban, and the social as well as the spatial. Jackson (1992) contends that the main thrust of new cultural geography is a study of the plurality of lived cultures, and of the multiplicity of landscapes and meaning that result. Hence, major themes of study are gender and sexuality, with research exploring the inter-relation of people, culture and place.

Conclusions

The conclusion should sum up the argument but without directly repeating statements from the introduction. Introducing new material can cause examiners to comment 'The conclusions had little to do with the text'. It is hard to judge the standard of conclusions in isolation, because the nature and style of a conclusion depend on the content of the essay, but it helps to look at other examples. I would advocate reading articles in *The Economist*. There is something relevant for physical and human geographers in most issues. Despite this caveat criticise the examples in **Try This 12.3**.

TRY THIS 12.3 – Good concluding paragraphs

Read the following concluding paragraphs and consider their relative merits. Comments on p.261.

Version 1

It is therefore possible to see that people have had a mega impact on the landscape as the land devoted to agriculture quadrupled and there was an explosion in urban development. This impact was caused by the discovery that wetlands have many important values and perform many worthwhile functions. The components of the system are seen to be interconnected, and that destruction of one part of the system will have effects, probably adverse elsewhere in the system. This has meant that for the last 25 years there has been less damage, but only a great deal of time and conservation will reverse the extraordinary exploitation that has occurred on Canada's wetland landscapes.

Version 2

All in all, there has been a lot of pressure on Canadian wetlands over the past hundred years due to human impact, but things look set to improve, with talk, and actions, of restoring many wetlands areas, and also new international regulations and laws for their protection.

Version 3

In addition to the impacts I have considered, threats likely to continue or increase include the harvesting of black spruce (*Picea mariana*), use of wetlands for wastewater treatment, burning which may lead to the loss to native species and the invasion of Eurasian weed species, and increased accumulation of sulphur and heavy metals from acid rain. The impact of people has led to the loss, or serious degradation, of these vulnerable ecosystems which are effectively a non-renewable resource. The recreation and educational value of wetlands has been discussed. There is a dilemma for managers here, who need to protect the ecology of sites while giving appropriate access for eco-tourists, hikers, ornithologists, fishermen and hunters. Good *et al.* (1978) pointed out that much wetland management derived from common sense rather than science. In 1999 the science base has increased, the complexities of ecosystems interactions are better understood. The RAMSAR Convention has provided valuable guidelines for wetland preservation but there is still much to be done to prevent further wetland destruction.

Version 4

Canadian wetlands were drastically altered by the intervention of humans. Seventy per cent have been reclaimed for a variety of purposes including agriculture, urban development, industry, energy development and harvesting. There is a consequent loss of habitat and reduced species diversity. Recent changes in human use of wetlands have led to some abandoned agricultural sites being returned to wetland but agriculture is still the dominant activity.

Version 5

To conclude, I think it important to highlight the damage done by people to wetlands, both on a national scale and at a local scale where only small amounts of interference can cause potentially devastating effects. This essay has shown that without greater management of these environmentally sensitive areas we will not only destroy a vast educational resource but also affect the entire global atmospheric system.

12.6 Assessment and Plagiarism

Assessment

Either use your department's assessment criteria or Table 12.1 and Figure 12.2 to critically review your writing. In Figure 12.2 the percentage equivalents are given for guidance, but these values vary. Self-assessing a first draft can indicate where to focus your next research and writing effort.

Another way of developing your evaluation skills is to use **Try This 12.4** which builds on an idea from note-making by using codes to locate the different sections of an essay and to compare their relative weights. **Try This 12.5** presents another way of creating essay structures by posing questions about the mark distribution. If the balance of the reward changes then so does the content.

Criterion	First 70–100%	2(i) 60–69%	2(ii) 50–59%	Third 40–49%	Ordinary 37–39%	Fail 0–36%
Structure: Organization, logical order of material, aims, conclusion	Well organized throughout. Good structure. Appropriate aims and conclusion specified clearly.	Mostly well organized, appropriate structure. Aims and conclusion specified.	Structure attempted. Evidence of organization. Sound aims and conclusion.	Some attempt at structure or organization but inappropriate.	Little attempt at organization.	Disorganized. No structure, aims or conclusions.
Accuracy and understanding of material, focusing on the question	Accurate and thorough understanding focused on the question throughout.	Good understanding of the subject, mostly focused on the question.	Sound understanding of the subject but not effectively focused on the question. Some inaccuracies or misunderstandings.	Inaccuracies evident. Some limited understanding of the subject, not applied to the question.	Some glimpses of understanding but much work inaccurate or unfocused.	Totally fails to address the question posed. Fails to demonstrate understanding of the subject.
Coverage: Comprehensive-ness, relevance, evidence of reading, research, use of examples	Developed own ideas based on thorough understanding of the relevant literature. Comprehensive coverage of material. Excellent use of illustrative examples.	Demonstrates evidence of reading well beyond lecture material. References are relevant. Good use of illustrative examples.	Covers lecture material reasonably well. Mostly accurate and comprehensive. Some evidence of further reading. Some relevant examples provided.	Lecture material only, but not comprehensive. Little evidence of further reading. Some examples given but not all relevant.	Much irrelevant or missing material. Too brief but some knowledge of the general topic. Little attempt to exemplify.	Lack of relevant material. No use of examples.
Clarity of argument: Coherent, fluent, criticality, innovation	Excellent coherence and clarity of expression. Demonstrates application of critical thought. Shows innovation in handling arguments.	Mostly coherent. Some attempt at critical analysis and innovative argument.	Generally coherent but some lack of clarity of thought or expression.	Some clarity but too simplistic.	Somewhat disjointed and lacking in development.	Incoherent. Disjointed.
Presentation: Grammar, spelling, legibility, referencing system	No errors. Clear, relevant and consistently accurate referencing.	Few errors. Referencing relevant and mostly accurate.	Occasional errors. Minor inaccuracies or inconsistencies in referencing.	Frequent errors. Referencing with some inaccuracies or inconsistencies.	Very frequent errors. Referencing inaccurate or inconsistent.	Riddled with errors. Referencing absent.

Table 12.1 Essay assessment criterion

Essay Title......						
	I	2.1	2.2	3	Fail	
Knowledge 30%						
Topics covered in depth						Superficial responses
Appropriate geographical content						Limited/no geographical content
Structure and Argument 40%						
Logical Presentation						Disorganised
Good synthesis and evaluation						No synthesis and evaluation
Clear, succinct writing style						Rambling and/or repetitious
Creativity 20%						
Includes new ideas						No new ideas
Innovative presentation						Incoherent presentation
Presentation 10%						
Fully and correctly referenced						No references
Correct spelling and grammar						Poor spelling and grammar
Good use of illustrative materials						Poor/no use of illustrative materials

Figure 12.2 Essay self-assessment form

TRY THIS 12.4 – Analysing essay structures

Look at any essay you have written. With 5 coloured pens, or a code system, mark in the margin the sections which show Knowledge, Analysis, Synthesis, Evaluation, and Creative abilities. Now think about the relative content and structure.

- How would you re-write this essay to increase the analysis section?
- How would you redesign to increase the evaluation content?
- Is there a good opening paragraph?
- Does the argument flow logically?
- Are the examples relevant?
- Are the arguments summarised effectively?
- Are the conclusions justified?
- Where could the balance of the essay be altered to improve it?

TRY THIS 12.5 – Marks for what?

Take any essay title (the last or next) and work out a plan for researching and answering the question when the mark distribution is:

1. Knowledge 20 per cent; Analysis 20 per cent; Synthesis 20 per cent; Evaluation 20 per cent; Professionalism 20 per cent,

AND

2. Knowledge 10 per cent; Analysis 50 per cent; Synthesis 20 per cent; Evaluation 10 per cent; Creativity 10 per cent,

AND

3. Knowledge 30 per cent; Evaluation 70 per cent

Consider how the plans change as the mark distributions alter.

Avoiding plagiarism

Academic staff are good at noticing words copied from texts and papers without acknowledgement. They spot changes in the style of writing, use of tenses, changes in formats and page sizes, the sudden appearance of very technical words or sentences that the writer doesn't seem happy with, and quotations without citations. The advice is 'never cut and paste any document', whether from the www or friends. Think through the point you want to make, express it in your own way and cite sources as you write. Put quotations in quotation marks and cite the source.

Getting advice

Showing a draft essay, or any document to someone else for comment is not cheating. It is normal practice in business and academic publishing, as shown by the acknowledgements at the end of many published papers. For example, in *Progress in Human Geography* 22, 1, 1998, there are thanks for reviewer's comments and suggestions on pages 13, 36, 50, and 70. You do not have to take note of the comments, but having an independent check on grammar, spelling and someone asking awkward questions about content does no harm.

12.7 GETTING THE ENGLISH ~~RIGHT~~ BETTER

Try ~~NEVER~~ to cross anything out! *Keep sentences short* wherever possible. See **Try This 12.6** for starters!

TRY THIS 12.6 – Shorten these sentences

Answers on p.262.

Wordy	Better
In many cases, the tourists were overcharged.	
Microbes are an important factor in soil processes.	
It is rarely the case that sampling is too detailed.	
The headman was the proud possessor of much of the land in the vicinity of the village.	
Moving to another phase of the project	
The flow rates in the Severn were monitored in August and October respectively.	
Chi-square is a kind of statistical test. (should read 'Chi-square is a type of statistical test')	
One of the best ways of tackling prison reform is ...	
The investigation of African cultural impacts continues along the lines outlined.	
The nature of the problem ...	
Temperature is increasingly important in snowmelt.	
One prominent feature of the landscape was the narrow valleys.	
It is sort of understood that ...	
It is difficult enough to learn about pollen analysis without time constraints adding to the pressure.	
The body of evidence is in favour of ...	

Technical terms

Good professional writing in any subject uses technical language, defining technical terms when necessary. Definitions can be very important, housing policies in Leicester are rather different from those in São Paulo, an intense rainstorm in Singapore is considerably more intense than one in Leeds, and a socialist or conservative government means different things under different regimes. There is no need to define technical terms that are in everyday use and where you are using words in their normal sense; assume the reader is intelligent and well educated. In an essay on hydrological flows in river basins, paragraph 2 started with: 'Water covers ⅓ of the earth's surface, mostly as oceans. Oceans are expanses of saline water which vary in dimensions, depth and width, and form e.g.: calm, rough, warm, cold'. In my view this student wasted time on the second sentence, an examiner knows what an ocean is, and it was not appropriate to define it in the context of that essay. Keep the language suitably technical and the sentences simple. Use the correct technical terms wherever possible and avoid being unnecessarily long-winded, as in 'snowmelt is largely a seasonal phenomenon, which occurs when temperatures rise after frozen precipitation has fallen'.

Startling imagery

'It is a truth, universally acknowledged, that a man in possession of a large screen satellite TV on Cup Final Day, must be in want of a six pack.' The impact is greater for being the antithesis of Jane Austen. However, this comment, while arresting, is also dangerously stereotypical, far from politically correct and genderist. Creative use of language is great, but avoid the temptation to bowdlerise or tabloidise in an inappropriate manner.

One idea per paragraph

In a lengthy essay, restricting paragraphs to one idea plus its supporting argument should make your message clearer. Use **Try This 12.7** to analyse one of your own essays.

TRY THIS 12.7 – One idea per paragraph

Look back through your last essay and underline the ideas and supporting statements. If each paragraph has a separate idea and evidence, award yourself a chocolate bar and cheer. If not, redesign a couple of paragraphs to disentangle the arguments and evidence. The challenge is to make the evidence clear to the reader by separating out the different strands of the argument.

N.B. This is a good exercise to do at the end of the first draft of every essay.

Synonyms

Part of the richness of English comes from the many synonyms that add variety, depth and readability to writing. Take care you write what you mean to write! A common error is to confuse infer and imply, they are not synonyms; infer is used when drawing a conclusion from data or other information, imply means to suggest or indicate. We might infer from a dataset that the world is flat, or a lecturer might imply during a discussion that the world is flat. Have a practice with **Try This 12.8** and **Try This 12.9**. Look at any sentence you have written, and play around with a thesaurus (book or electronic) and find synonyms you might use. If you tend to over-use certain words, make a list of synonyms and substitute some of them, WHERE RELEVANT.

TRY THIS 12.8 – Synonyms for geoggers?

Which of the two synonyms makes geographical sense? Answers on p.263.

The sentences were extracted at random from *Progress in Human Geography* 21, 3, 1997 and *Progress in Physical Geography* 21, 2, 1997.

1. Studies on litter and organic matter dynamics also assist awareness/understanding of nutrient mineralization.
2. I am working on a research project that uses in-depth interviews as its main/paramount data source.
3. The discrimination/distinction and mapping of rock types are a major focus of geological remote sensing.
4. River networks can be deduced/extracted from satellite images allowing controls on drainage patterns to be analysed.
5. What economic logic elucidates/explains plant closings and firms' exit from industries?
6. Quantification is particularly important if real appraisals/estimates of the extent and timing of hazard impacts are to be useful for forecasters.
7. Past experience/practice of negotiating emission agreements suggests that despite legislation, success in emission control is by no means assured.
8. Any process by which a marine-ice sheet may form must overcome the problem of enhanced calving rates associated/comparable with deeper water.
9. There is some limited documentation of the effects of channel and water management on the morphology/shape of channels in the Mediterranean region.
10. At one level Dodd's work can be seen as sympathetic to that of Zelizer, for he is similarly critical of the way in which money has been treated in traditional/conservative social theory.

TRY THIS 12.9 – More synonyms

These two grids show a perfectly acceptable sentence and some of the synonyms that might be substituted. Some synonyms are acceptable, others make no sense at all. Which are useful substitutes?

			society						
	study	including	family		assistants			decent	scope
Any	inquiry	comprising	humanity	because	contributors			equitable	range
Each	examination	containing	tribes	during	helpers	owns		worthy	compass
All	*research*	*involving*	*people*	*as*	*participants*	*has*	*an*	*ethical*	*dimension*
Every	investigation	entailing	nations	while	associates		a	just	sweep
Whole	scrutiny	embracing	races		collaborators			upright	limit
		encompassing	groups		colleagues				size

Countryside		Duration			thesis's	primary	ideas
Vistas		Past			argument's	main	notions
Perspective		Future			content's	foremost	thoughts
Landscape	*and*	*Time*	*are*	*the*	*subject's*	*key*	*concepts*
Panorama		Span			topic's	index	inspirations
Scenery		Cycle			point's	catalogue	images
Environment		Season			substance's	first	impressions
						leading	

Argument by analogy can be very useful, but this example shows it can also get out of hand*:* 'A fast breeder reactor is more efficient than a normal nuclear reactor. It can be thought of as a pig, eating anything, rather than a selective racehorse.' In this case the first sentence would be better on its own.

I know what I meant! Trying to get students to re-read and correct written work is an uphill battle for most tutors. Re-read to spot illogical statements or those where a crucial word was missed. Here for your amusement are some geography essay statements that do not exactly convey the writer's original intention: 'For centuries cars have been causing problems on the roads' (when were they invented?); 'No data can be gathered as the area is relatively flat'; 'Large-scale windmills should be located in areas where winds blow almost inconsistently'; 'when it's really dry the soil gets to wilting point and the plants become extinct'; 'A random model has lots of strange processes usually found by throwing dice. To get a weighted random model you need a weighted dice'; 'Blackpool has made concreted efforts to manage tourists.'

Abbreviations and acronyms in text

Replacing long words or phrases with initials or abbreviations is regarded as lazy by some tutors. However, with well-known and established acronyms (see p.248) and phrases like NIMBY (Not in my back yard), it is reasonable to adopt this approach. Consider your response to abbreviations in **Try This 12.10**. When using abbreviations, the full definition must be given the first time the phrase is used with the abbreviation immediately afterwards in brackets. Biogeographers and botanists conventionally adopt an abbreviation strategy when citing the Latin names of plants (see p.240).

TRY THIS 12.10 – Abbreviations?

Look at Chapter 17. How do you feel about the use of ES as an abbreviation throughout this chapter? Is it a procedure to adopt?

Colloquial usage

Regional or colloquial terms may not be universally understood so are best omitted. Writing as you speak can also be a trap, as in these examples from student essays:

'He therefore put to greater emphasis on the results of . . .' It should read 'He therefore put too great an emphasis on . . .'
'The large events will of always penetrated through the pervious ground', rather than 'the large events will have always penetrated through . . .'.

Punctuation and spelling

The excessive use of exclamation marks!!!!!!!!!!, of which this author is generally guilty, is also less than good practice! *And never* start a sentence with And, But or BUT. There are plenty of examples of its incorrect usage in this book, where BUT and capitalisation are used to emphasise points.

Spelling is potentially a minefield, use the spell checker but remember that it will not pick up the errors in the following sentences: 'the Haddock shopping centre' (in this case, Haydock was the example), 'In general, models of systems can be said to be lumped or disintegrated', 'focused on the shear pace of change', 'middlemen can charge exuberant prices', 'the Department of Fare Trading', 'the competition between predators and pray', 'The site (near Halifax, West Yorkshire) is upstream of a fjord to a residential property.'

When editing, check that you do not over-use certain words. Find synonyms or restructure the paragraph if repetition is a problem! Keep sentences short and TO THE POINT. Ensure paragraphs address one point only. Be consistent in your use of fonts and font sizes, symbols, heading titles and position, bullet points and referencing. Decide on your style and stick to it. If you feel you have trouble with your writing, or you are unsure about the use of colons, semi-colons and apostrophes see Kahn (1991), Kirkman (1993) or Russell (1993).

Top Tips

- Read and revise everything you write. Make time at the end of an essay to reread and redraft, correct spelling, insert missing words, check grammar, tidy up diagrams and insert references.
- Check that your arguments are logical.
- Read what is written, not what you meant to write.
- Work with a friend.

12.8 References and Further Reading

Some genuine references were embedded in some of the essay extracts, and the full citation would appear at the end of the essay. They are not included here. A BIDS search will find them, or see
http://www.geog.leeds.ac.uk/staff/p.kneale/skillbook.html

For grammar queries see:

Dummett, M. 1993 *Grammar and Style for Examination Candidates and Others*, Duckworth, London.

Kahn, J.E. (ed.) 1991 *How to Write and Speak Better*, Reader's Digest, London.

Kirkman, J. 1993 *Full Marks: advice on punctuation for scientific and technical writing*, (2nd edn) Ramsbury Books, Marlborough.

Most study skills books discuss essay writing, but see also:

Barrass, R. 1982 *Students Must Write – a Guide to Better Writing in Course Work and Examinations*, Routledge, London.

Fairbairn, G.J. and Winch, C. 1996 *Reading, Writing and Reasoning: a guide for students*, (2nd edn) Open University Press, Milton Keynes.

Fitzgerald, M. 1994 Why Write Essays? *Journal of Geography in Higher Education*, Directions, **18**, 3, 379–84.

Lillis, T. 1997 Essay Writing Starter, in Drew, S. and Bingham, R. (eds) *The Student Skills Guide*, Gower, Aldershot, pp. 53–75.

Russell, S. 1993 *Grammar, Structure and Style*, Oxford University Press, Oxford.

Geograms 3

Try these geographical anagrams. Answers on p.263.

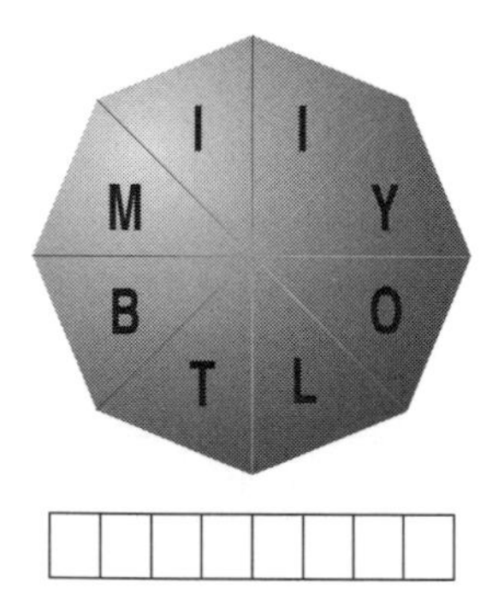

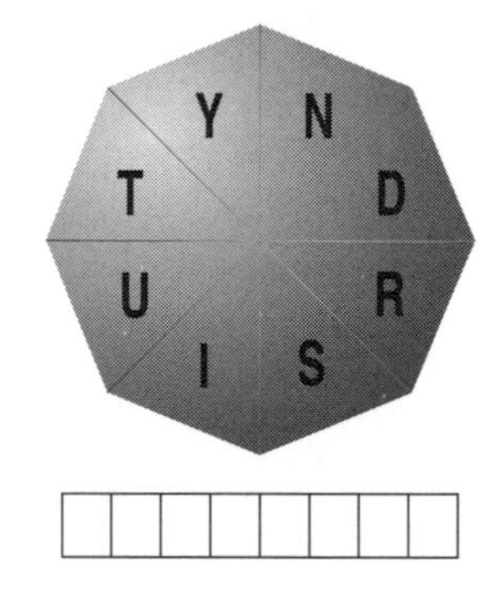

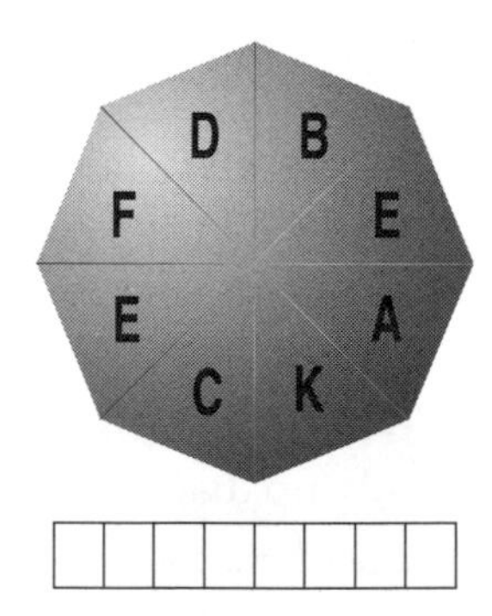

13 RESEARCHING AND WRITING IN TEAMS: IT'S FUN AND EFFICIENT

How do you know what you think before you talk about it?

Employers look for people who work happily in teams. Soloists may be very useful, but if s/he cannot get on with colleagues, explain and relax in management situations, that person may impede progress. More importantly, teamwork allows the exploration of more material than is possible for an individual. Team members can bounce ideas off each other. Where resources are limited, getting involved in co-operative activities and sharing has benefits for everyone. The skill benefits include teamwork, developing professional standards in presentation, problem solving, negotiation and responsibility.

A bunch of geographers having supper together may talk about Prof. Blownitts' lecture on nuclear power. Everyone has different ideas. If some people decide to chase up further information and pool it, they are beginning to operate as a team. In some modules you will research and write as a team, but remember you can use team skills and approaches to tackle other parts of your course. Two to five people in the same flat, hall of residence or tutor group can team up to extend and optimise research activities, and talking geography together will develop your discussion skills and broaden your views.

This chapter raises some of the issues associated with group work and suggests some ways of tackling group tasks. There are pitfalls. Not everyone likes teamwork and sharing. Some people feel they may be led down blind alleys by their team-mates. Overall, I think the advantages of team research outweigh the disadvantages, but if this is an issue that bothers you, brainstorm reasons why certain group members are a disadvantage. Possibly:

- ☹ those who do not pull their weight;
- ☹ the perennially absent;
- ☹ those who do not deliver on time;
- ☹ the over-critical who put others down, suppressing the flow of ideas;
- ☹ anyone who gets touchy and sulks if their ideas are ignored.

If you are aiming to benefit from team activities, these are some characteristics to suppress! You may not like working in teams, but when you do, it is in everybody's interests for the team to pull together and derive the benefits. Having identified unhelpful teamwork characteristics, consider what good team qualities might include:

☺ making people laugh;

☺ getting stuff finished;

☺ communicating well;

☺ being reliable;

☺ making friends;

☺ keeping calm in arguments;

☺ speaking your mind;

☺ resolving disputes.

Reducing tension in a group promotes cohesion and encourages learning.

There is a natural pattern in life that can also be seen in the normal reaction of people to teamwork and life in general (see Fig. 13.1). Map these natural reactions to an event in your life, or the response of the Conservative Party after its defeat in 1997, or to an assignment as in Fig. 13.1. Each time you are given an assignment you are likely to experience all these reactions. The people who realise 'where' they are in the sequence, and push on to the Recovery and Getting on Track stages, give themselves more time to do the task well.

Natural Reactions	Reactions mapped to an Essay	Reactions mapped to a Team Project
Shock	I know nothing about ...	We know nothing about ...
Recrimination	I didn't want to do this module, who made it compulsory?	I don't want to work in groups/ with them.
Disagreement	No motivation, nothing done.	Everyone does their own thing, or nothing.
Reorganisation	2 weeks to go, do a plan.	There are 2 weeks to go, we have to get it together! Make a plan.
Recovery	Parts of this are quite interesting.	Everyone knows which bits to do. Some people start to enjoy it.
Getting on track	Got parts of two sections sorted, and found some more references.	It is coming together, disks are swapped and ideas evaluated.
Partial success	Drafted the essay, needs editing and references.	Mostly done, but no cover, diagrams missing and needs editing.
Frustration	Printers not working, forgot to spell check.	We can't find Jim who has it all on disk!
Success	Final version handed in.	Report completed.

Figure 13.1 Natural reactions to group work

13.1 Researching in Groups

Groups can cover more material than an individual, but having decided to share research outcomes, recognise there will be difficulties and tensions at times. Group members are only human. Everyone has a different way of researching and note making, so shared information will 'look' different on paper, BUT you can learn from the way other people note and present information. Figure 13.2 summarises thoughts from three groups after a shared research task. They give an insight into what can happen.

One meeting or three? How often does a group need to meet? Generally two or more meetings ➜ more chat ➜ greater exchange of views ➜ more interactions ➜more learning. Have electronic meetings (see p.141) when time is tight.

13.2 Writing in Teams

Team writing is a normal business activity. Typically it involves a group brainstorming an outline for the document, individuals researching and drafting sub-sections, circulation of drafts for comment, the incorporation of additional ideas and views, and someone editing a final version. The same approach can be adopted in university writing. Like ensemble recorder playing, writing as a team is difficult and discordant the first time you do it, but very useful experience.

Writing Role	Responsibilities
Summariser	Introduction, Conclusions, Abstract
Visuals/Graphics	Smart graphs, figures, indexes, contents page
Critic	Faultfinder, plays devil's advocate
Academic Content	Subject reporters
Linker	Checks connections between arguments, sections and the introduction and conclusion
Discussant	Evaluates and discusses

Figure 13.3 Writing roles

One valuable approach involves team members volunteering to research and draft specific sub-sections and volunteering for a 'writing role'. Discussions are more lively and focused when individuals can use their role to offer alternatives and new approaches. The critic has the licence to criticise, the linker can say, 'Yes that's great, but how do we use butter mountains and polar ecology to explain county employment patterns?' Adopting writing roles can depersonalise criticism, which is especially important when working with friends. Be critical and stay friends.

Task	Reflection
Tutorial Essay on Wind Power Individual essays Two-week deadline	*Met and decided we didn't know where to get data! Split up to do library, BIDS and www search. Discovered power company brochures and reports. Agreed to keep 2 texts and photocopies in Ed's kitchen cupboard where we could all get at them. Middle of week 2 we realised no one had looked at Wind Power alternatives, Sarah did it. Two discussions at lunchtime in the Union, and it became the topic of conversation in Ed's kitchen for a week. Sarah and Mike phoned two companies for further details and got some info, but too late for the company to post things to us before the deadline. The research time was about equally split. Generated a bigger pile of articles at Ed's than we could have found alone. Results were a bit general. We needed a better plan at the start or part way through week 1. Found info in the Engineering Library at the end, I think we thought it would be too difficult for geographers, but it would have helped. We got lots of electronic source info, which saved time, BUT it cost a lot of printer credits. In future we should be more selective or e-mail to each other to save printing. Good to chat, group got to know each other better. One person thought s/he would have started writing sooner, but waited for info from others.*
Seminar Preparation Four people Group presentation 20 minutes followed by general discussion, class of 25	*We all knew each other and it was easy to share out the reading, we volunteered to cover different sub-topics. We decided to meet 5 days before the seminar, over supper, to discuss who was doing what. Great evening, great discussion, no one wrote anything down, so we had to meet the next day to draft the talk.* *We had far too much material but we could choose. Took examples from Bosnia (it's topical) and Sudan (map and pictures available). Put other refs on a handout. Decided there were some good www sites, but no interactive computer display system available for the presentation, so we put web site details on the handout. Decided to use an 'awkward questions' format. Put awkward questions on OHT and then decided who would try to answer each one. Changed the questions a bit! Met the day before to practise.* *Everyone talked for too long, tried to cut bits out. The introduction changed about twenty-five times! The seminar went OK, mostly because we had talked about the points beforehand, so I think we knew more than usual.*
Organising a field research day for social geography Five people Task to be decided	*Sorting out what and how to do it took ages, there were too many good ideas. We had trouble getting together, and in finalising the plan. Tended to split into two. The organisation of the day was left to Andrea in the end, because she had the car. We met three times beforehand and didn't really decide anything each time. We needed a leader from the start and to focus faster on what to do.* *The field day was great FUN, Liz and Jamie did a questionnaire on her pc, we changed it about a bit over lunch. Didn't get as much data as we hoped, but talked to some really nice people and the graphs look good. We had loads to write under the 'criticism' and 'what we would do next time' sections.*

Figure 13:2 Experiences of group research

Style and layout

Decide on the general layout at the start, on the format, fonts and style of headings, and the format for references. Make decisions about word length, like 'each subsection has a maximum of 200 words'. Early word length decisions limit waffle. Nevertheless, length needs looking at later in the project, as some parts of the argument will deserve expansion. Keep talking to each other about ideas and their relative importance and position in the narrative.

Finally, someone (or some two) has to take the whole document and edit it to give it a consistent voice and style, BUT everyone needs to provide graphs, diagrams and references in the agreed style.

Duet writing

Two people sitting at a pc can be very effective in getting words on disk. The exchange of ideas is immediate, two brains keep the enthusiasm levels high and you can plan further research activities as you go. It is advantageous to have two people together doing the final edit, keeping track of formats and updates and keeping cheerful.

Timetabling issues

A team writing activity cannot be done like the traditional geography essay, on the night before submission after the pub. People get ill and things happen, so the timetable needs to be generous to allow for slippage AND team members have to agree to stick to it. It usually works, as it is too embarrassing to be the only non-contributor. It may assist planning if you put some dates against the points as shown here:

1.	Brainstorm initial ideas, assign research tasks and data collection *(Day 1)*
2.	Research topics *(Day 2–5)*
3.	Draft subsections and circulate *(Day 6–9)*
4.	Meet to discuss progress, decide on areas that are complete, assess where additional research is needed, assign further research and writing roles and tasks *(Day 10)*
5.	Redraft sections and circulate *(Day 10–14)*
6.	Are we all happy with this? *(Day 15)*
7.	Editor's final revisions *(Day 16–18)*
8.	Finalisation of cover, contents and abstract. Check and add page numbers. Check submission requirements are met. *(Day 18)*
9.	Submit *(Day 18!)*

Keeping a team on track? Need a chairperson?

One of the pitfalls of working with a group of friends is that more time is taken making sure everyone stays friends, than getting on with the task. One vital issue that emerges is the initial division of labour (see Fig. 13.2 and Fig. 13.5). It is vital that everyone feels happy, involved and equally valued. So the chairman, or the person that emerges as the chairman, must endeavour to ensure there is fair play, that no-one hogs the action excluding others, and equally that no-one is left out (even if that is what they want.) The chairman is allowed to goad you into action, that's her/his job. It is unfair to dump the role of chairman on the same person each time, share it around. Chairing is a skill everyone should acquire.

If a group feels someone is a serious dosser they may want to invoke the 'football rules', or ask the module tutor to do so. The rules are one yellow card as a warning, two or three yellow cards equal a red card and exclusion from the group. The yellow–red card system might be used to reduce marks (see p.142). A red carded person does not necessarily get a zero but in attempting to complete a group task alone, they are unlikely to do better than a bare pass.

- ✔ The time required to tidy up, write an abstract, make a smart cover, do an index, acknowledgement page and cross-check the references is five times longer than you think, so split the jobs between the team.
- ✔ It saves hours of work if everyone agrees at the start on a common format for references, and everyone takes responsibility for citing the items they quote.

The key to team writing is getting the STYLE and TIMETABLE right, although CONTENT also matters.

Plagiarism

Where group work leads to a common report, then obviously collaborative writing is involved. BUT, if you are required to write independent reports from group research or group activities you must ensure your reports are independently written, not copies or cut and paste versions of each other's documents. In this situation share reading, float and discuss ideas, BUT write independently. This means planning to finish the research three days, or more, before the deadline so everyone has time to draft and correct their reports.

13.3 Communication by e-mail

It is vital that everyone has access to a group report, therefore a method of circulating the most recent version of a document is needed. Using e-mail with attachments is ideal, and saves printing costs. You can share thoughts, drafts and updates, while working at the most convenient place and time for you. Have a go at **Try This 13.1** as practice.

TRY THIS 13.1 – E-mail in action

Research a tutorial essay or practical report, sharing resources and ideas, without physically meeting. HINT: the first time, it helps to have everyone in the same computer laboratory, you can resolve any problems quickly.

Open a new word processing document and:

1. Type in the title, the keywords you would use in a library search and write very briefly about two or three issues, say, three sentences for each.
2. Then save the document. Open e-mail, attach and mail the document to your mates.
3. With a bit of luck other people will be doing the same thing at the same time, so there should be e-mails arriving from them. Save your colleagues' files in your own workspace, but check whether they have used the same file name as you, if so it will need changing to avoid over-writing.
4. Return to the word processing package, open your original document and the new ones. Copy the material into a single file and re-organise it to make a coherent set of comments. While collating keep track of ideas, perhaps through sub-headings. Reflect on your colleagues' comments. At the immediate supportive level there is the 'that's a new/good/middling idea'. More actively, think around variations on 'I was surprised by ... because ...' 'I disagree with ...because ...' 'We all agreed that ... because ...'.
5. Now look at the *style* of the responses. What might you need to do as an editor to make these comments hang together?
6. Finally, get together and decide how you will organise your research and writing to maximise the opportunity to share resources and information.

Bigger documents

Imagine a report constructed by four people, being edited and updated daily. How do you keep track of what is going on? There needs to be an agreed system. Adapting elements of these Top Tips might be a useful template for action:

- The start of the document needs a:
  - header page with title and the outline plan;
  - section which everyone updates when they change the document e.g. Andy modified section 2.6 and 3.2 on Wed 1 April at 10.00;
  - an agreed working order, e.g. John drafts section 2, then Anne revises it. All revisions circulated by … day.
- Decide that citations will be added to the reference list in the format agreed.
- Agree to check e-mail and respond every X days.
- Use the 'Revision Editor' in your word processing package. The person responsible for section X will not necessarily welcome five independent redrafts. A 'Revisions Editor' highlights revision suggestions for the subsection editor to accept or reject as desired. By opening multiple copies of the document you can cut and paste between drafts.
- DO NOT USE PAGE NUMBERS TO REFER TO OTHER VERSIONS. These change with almost every version. Use section numbers and date each version carefully.
- Ensure at least one or two people keep an archive of all drafts, so that you can revive an earlier version if disaster strikes.

Use e-mail to brainstorm ideas among friends, tutorial and seminar groups and old school friends doing geography at other universities. People like getting messages and usually respond. E-mail is a quick and cheap way of discussing points. There are a great many people you could brainstorm with, think more widely than just you and your lecturer.

13.4 Assessment of team work

Assessment tends to generate discussion about 'fairness'. There are dark hints about cheerful dossers getting good marks when their mates have done the work. How is this handled? Most staff will offer some variation on the following approaches for assigning marks, some of which involve team input. If the 'football factor' is in play, there may be an agreed penalty, say –10 per cent for a yellow card.

The simple approach

Each member of the team gets the same mark, so it is up to everyone to play fair.

The private bid

Each individual fills in a form privately, and the assessor tries to resolve discrepancies. This style:

Name ....
Names of other Team Members ....
I feel my contribution to this project is worth ....% of the team mark.
This is because ....
....
....
Signed ....

This approach may lead to discussion amongst your assessors, but it lets people with personal problems acquaint staff privately. It can require the wisdom of Solomon to resolve. Remember a marker will take note of what you say, but not necessarily change the marks.

Team effort

A form in this style asks everyone to comment on the contribution of each team member. Summing the totals constructs an index of activity, which is used to proportion the marks.

Estimate the effort made by each team member, 0 = no effort, 2 = did a bit, 3 = average, 4 = really useful, 5 = outstanding contribution.	Tracey	Steve	......N
Attended all meetings			
Contributed ideas			
Did a fair share of the writing			
Other – detail particular contributions:			
TOTAL			

13.5 COMMENTS ON TEAM WORK

Staff comments: 'Initially students are very democratic and give equal marks. With experience, they choose to raise the marks of those who have done more of the task'. 'The amount of chat was enormous, and between them they could tackle a more difficult problem than as individuals'. 'The team writing exercise made everyone think about the order and quality of the material. The report was more extensive and detailed than one person could do in the time'. 'One team was happy to have this year's 'mine's a third candidate' because s/he happily did the activities like washing up in the laboratory and making poster backgrounds, while cracking jokes and keeping the team cheerful'.

Figure 13.4 details reflections of first year students following group research using library and electronic sources. Overall, they see the exercise as positive and the disadvantages as surmountable with experience. Their reflections may be worth considering as you start team activities.

What were the advantages of team work?	**What were the disadvantages of team work?**
Shared ideas. Divided workload saved time. *New Friends/Met new people.* *Pulling ideas from different viewpoints.* *Less work per person. People can help each other to understand. Learn consideration.* *Lots of prior knowledge got us started quickly.* *Created cohesion amongst 5 people who had only just met. Greater overview of the subject.* *We covered library, www, CD-ROM and E-journals between us, more than we would have managed alone. Learning to share and compromise.* *Individuals could concentrate on the bits that interested them. Encouraging to know that others in the team thought you were on the right lines.*	*Conflicting views (is this an advantage?).* *Having to compromise on common topics and ideas.* *Hard to get a time to meet.* *Difficult to agree on points of view.* *Who does the writing?* *Hard to find out if a non-contributor was ill or skiving (was ill).* *Recognising that other people have different ways of working, and having to find a way round it!*

What will you do differently next time you work together?
More thorough planning to concentrate efforts/Set clear objectives at the start. *Improve web skills to save time/e-mail each other/get used to reading e-mail regularly.* *Elect a team co-ordinator, who has the team members' timetables, to find convenient meeting times.* *Split the team to tackle different libraries and meet later.* *Organise a couple of meetings rather than trying to do it all in one go.* *Reserve books in advance/arrange to meet in library at a quieter time of day/search electronic sources first, papers and books later.* *Co-ordinate reading so two people do not read the same text.* *Start work on the project earlier/Not leave doing this to the last three days.* *Distribute the workload earlier.*

Figure: 13.4 Geography students' opinions on team work, 1997

Team research and writing at university usually produce a better document than an individual response. This is not because the academic content is necessarily substantially better, but because team activities generate a series of drafts and more thinking about the topic and audience, so that the final product is more polished. Make the most of group work activities on your CV. It isn't the academic content that matters, highlight the skills you used in delivering a group product, like negotiation, meeting deadlines, allocating tasks, collaborative writing, editing and co-ordinating research.

13.6 References and Further Reading

Anon, 1995 How to Build Effective Teams, *People Management*, 23 February, 40–1.

Belbin, R.M. 1996 *Management Teams: why they succeed or fail*, Butterworth Heinemann, Oxford.

Gibbs, G. 1994 *Learning in Teams: a student manual*, Oxford Centre for Staff Development, Oxford.

Vujakovic, P., Livingstone, I. and Mills, C. 1994 Why Work in Groups? *Journal of Geography in Higher Education*, Directions, **18**, 1, 124–7.

Why does a Geography professor sit the wrong way round in class?

Because she knows her subject backwards

14 ACKNOWLEDGING REFERENCES AND OTHER SOURCES

Describe the universe and cite the evidence for five comparable examples.

At the end of every piece of geographical writing you MUST include a reference list. This is an alphabetical list of all the sources actually quoted in your document, whether you read them or not. A reference list does not include 'other things I read but didn't quote in the text'. You may be asked to produce a bibliography, either in addition to your reference list, or as a task in its own right. A bibliography is an alphabetical list of sources or references on a particular topic; a complete bibliography would include every document relating to a topic. To create an annotated bibliography, sort the references into subsections with a brief statement or paragraph justifying your groupings and describing the contents.

There are a number of standard ways to acknowledge research sources. Some geography departments have a preferred style, so check the student handbook, or follow the advice here. These recommendations follow a standard pattern and adopt conventions cited by Xia and Crane (1996) and Cross and Towle (1996). The skill with referencing is consistency. Decide on a style and stick to it.

14.1 Citing paper sources in text

Within text, a book or article is cited by the author's family names and year of publication. When there are two authors both are quoted, but with three or more authors the *et al.* convention is adopted. For example 'Describing the retail geography of Great Buyit, D'Benham (2010) showed that the mega mall developments described by Markit and Hyper (2007) were already unprofitable, whereas the outlet distribution system analysed by McBurgers *et al.* (2008) had already expanded to ...'

Where information in one text refers to another quote both: 'As reported by Ward (2010), Parish (2008) found that ...' Both the Ward (2010) and Parish (2008) references should appear in the reference list. Similarly: 'In an extensive review of kluftkarren, Gryke (2010) shows the field approach taken by Clint (2005) is unreliable, and therefore the methodology adopted by Clint is not followed.' Quote both sources, although you have probably only read Gryke. Quoting both tells the reader how to locate the original. If you want to make clear that you have acquired your information from a secondary source use a sentence like 'Landslides in North America annually injure 5000 people and cause property damage in excess of $12 billion (Crushum 2005, cited Flattenem *et al.* 2010)'. In this case it is important to give both dates to indicate the age of the original data, 2005, rather than the 2010 date of the reference you read. You should quote both Crushum and Flattenem *et al.* in your references. The Crushum reference should be cited in the Flattenem *et al.* paper, so not including it would be lazy. If Flattenem *et al.* does not cite Crushum, use a BIDS search (see p.37) to find it.

Take care with oriental names where given names are second, the family name first. It is all too easy to reference by the given name.

Referring to government publications, where the author is awkward to trace, is also problematical. There are no absolute rules, use common sense or follow past practice. This example is a classic referencing nightmare:

CSICSC 1992 *China Statistical Yearbook* 1992, Fan Z., Fang J., Liu H., Wang Y. and Zhang J. (Eds.) China Statistical Information and Consultancy Service Centre, Beijing.

There is no single right way to cite this source, even librarians have different views on how to handle this one. Some would reference it by the editor as 'Fan *et al.* (1992)', others by the full title '*China Statistical Yearbook* (1992)'. In a library search you might have to try a number of options. Searching by title is likely to be the fastest successful route to locating this volume.

Referencing by initials can be convenient and time saving. You might cite DoETR (1998) or IEA (1997), but it is vital that the initials are explained in the reference list:

DoETR 1998 *Making Biodiversity Happen: a supplementary consultation paper to Opportunities for Change*, Department of the Environment, Transport and the Regions, London.

IEA (International Energy Agency) 1997 *International Coal Trade, the evolution of a global market*, Organisation of Economic Co-operation and Development, Paris.

Finally, if there doesn't seem to be a rule, invent one and use it consistently.

14.2 Citing paper sources in reference lists

The key is consistency in format, using a standard sequence of commas, stops and italics. Underline the italicised items in hand-written documents.

Citing a book

Template:
Author(s) Year *Title*, Edition, Publisher, Place of Publication.

Example:
Huggett, R.J. 1993 *Modelling the Human Impact on Nature: systems analysis of environmental problems*, Oxford University Press, Oxford.

Citing a chapter in an edited volume

The authors of the chapter or paper in an edited text are cited first, followed by the book editor's details. The title of the book is italicised, not the chapter title.

Template:
Author Year Chapter title. In Editors Name(s) (ed(s)) *Volume title*, Publisher, Place of Publication, Page Numbers.

Example:
Oke, T.R. 1997 Urban climates and global environmental change. In Thompson, R.D., and Perry, A. (eds) *Applied Climatology, principles and practice*, Routledge, London, 273–288.

Citing an edited book

Template:
Editor(s) (ed(s)) Year *Title*, Edition, Publisher, Place of Publication.

Examples:
Brooks, S. and Stoneman, R. (eds) 1997 *Conserving Bogs: the management handbook*, The Stationery Office, Edinburgh.

There are rare exceptions. There are three editors of this book, and a request to cite the RSPB, NRA and RSNC as authors:

RSPB, NRA and RSNC 1994 *The New Rivers and Wildlife Handbook*, The Royal Society for the Protection of Birds, Sandy, Bedfordshire.

Citing a journal article

Template:
Author Year Article Title, *Journal Title*, volume number, issue number, page numbers.

Example:
Dicken, P. and Miyamachi, Y. 1998 'From noodles to satellites': the changing geography of the Japanese sogo shosha, *Transactions of the Institute of British Geographers*, **23**, 1, 55–78.

Where no author attribution, use the Anon convention:

Anon, 1995 How to build effective teams, *People Management*, 23 February, 40–41.

Citing a newspaper article

Most newspaper articles have an author attribution, for example Fidler (1998). Where there is no author, use the first words of the headline as the cross-reference, and put the full headline in the reference list, as in Tractor Tragedy (1998).

Template when an author is cited:
Author Full Date Title, *Newspaper*, Volume number if applicable, Page Number(s).

Example:
Fidler, S. 12 March 1998 Feeling the chill winds from Asia, *Financial Times* Survey, Latin American Finance, 1.

Template for an unattributed item:
Title, Full Date *Newspaper*, Volume number if applicable, Page Number(s).

Example:
Tractor Tragedy Hits Village Community, 16 March 1998 *The Borchester Echo*, 1–3.

Citing an unpublished thesis

Thesis citations follow the general guidelines for a book (however the title is not generally italicised), then add 'unpublished', and enough information for another researcher to locate the volume.

Hoy, C.S. 1998 The Fertility and Migration Experience of Migrant Women in Beijing, China, unpublished PhD thesis, University of Leeds.

Holt, A. 1997 The Process of Designing a User-Friendly Environmental Assessment, unpublished BSc dissertation, School of Geography, University of Leeds.

14.3 Citing electronic sources

A standard template for citing electronic sources of information is not yet agreed. These notes follow recommendations from various library sources. If you are writing for a publication check whether an alternative method is used. The crucial new element is adding the date when you accessed the information, because the contents of electronic sites change. The next person to access the site may not see the same information.

Referencing within the text

Treat Internet and other electronic sources like printed reference. For example 'from maps of chemical pollution releases (Friends of the Earth 1997), we calculate . . .'.

Citing individual Internet sites

To cite Internet sources use the document's URL (Internet) address. Addresses tend to be long, so need careful checking. If the citation is longer than one line the URL should only be split after a forward slash/in the address. ThecaSe/ofchaRacters/inTheAddress/sHouldnOt/bealterEd.EVER

Template:

Author/editor, Year. *Title* [on-line] (Edition.) Place of publication, Publisher (if ascertainable), URL, Accessed Date

Example:

Friends of the Earth 1997 *Chemical Release Inventory of England and Wales,* [on-line] Friends of the Earth, UK, http://www.foe.co.uk/cri/html/indexmap.html Accessed 10 January 1999

When the electronic publication date is not stated write 'no date'. The term [on-line] indicates the type of publication medium. Use it for all Internet and e-journal sources. The 'Accessed date' is the date on which *you viewed or downloaded* the document.

'Publisher' covers both the traditional idea of a publisher of printed sources, and organisations responsible for maintaining sites on the Internet. Many Internet sites show the organisation maintaining the information, but not the text author. If in doubt, ascribe authorship to the smallest identifiable organisational unit.

Example:

Ordnance Survey 1998 *Mapping Britain's Health: Epidemiology, epidemiological study of asthma and road traffic in London*, [on-line], Ordnance Survey, Southampton, http://www.ordsvy.gov.uk/literatu/promo/epidemio.html Accessed 10 January 1999

Citing E-Journals (Electronic Journals)

Template

Author Year Title, *Journal Title,* [on-line], volume, issue, page numbers or location within host, URL, Accessed Date

Example:

Usunoff E. and Varni, M.R. 1995 Nitrate polluted groundwater at Azul, Argentina: characterisation and management issues, *Journal of Environmental Hydrology*, [on-line] 3, 2, 1-6. http://www.hydroweb.com/jeh_3_2/nitrate.html Accessed 10 January 1999.

In some e-journals the 'page' location is replaced by screen, paragraph or line numbers.

Citing personal electronic communications (e-mail)

To reference personal e-mail messages use the 'subject line' of the message as a title and include the full date. Remember to keep copies of the e-mails you reference.

Template:

Sender (Sender's E-mail address), Day Month Year. *Subject of Message*. E-mail to Recipient (Recipient's E-mail address.)

Example:

Bailey, D. (geo5467@Leeds.ac.uk), 21 October 1999. *Essay for Second Tutorial*, E-mail to P. Bradley (geo9876@Leeds.ac.uk.)

Top Tips

- **Authors** An author's name may be found at the foot of an electronic document. Authors of e-journal articles are usually cited at the beginning as in hard copy. Where the author is unclear, the URL should indicate the name of the institution responsible for the document. However, this organisation may only be maintaining the document, not producing it, so take care to assign the right authorship.
- **Date of publication** This is often at the foot of the page with the author's name, and sometimes with 'last updated' information. In newer versions of Netscape and other browsers, select *Document Info.* on the View menu. This shows the 'last modified date' of the document.

You must keep accurate records of the material you access. Using an on-line database bibliographic package can help to keep track of research resources.

Producing correct reference lists is an important skill, demonstrating your attention to detail and professionalism. Correct **Try This 14.1** to develop this skill. The ultimate test of a reference list is that someone else can use it to locate the documents. Check your citation lists meet this standard.

TRY THIS 14.1 – The nightmare reference list

There are many deliberate errors here. If a reference list like this appears at the end of an essay or dissertation, the marks will drift away. How many errors can you spot in 5 minutes? PLEASE DO NOT USE THIS LIST AS AN EXAMPLE OF GOOD PRACTICE!, use the corrected version on p.265.

Carver, S. and Openshaw, S. 1992, *A Geographic Information Systems approach to locating nuclear waste disposal sites*. In M. Clark, D. Smith, and A. Blowers eds Waste location: spatial aspects of waste management, hazards and disposal, Routledge, London, pp 105–127.

Abercrombie, N. and Warde, A. 1988. Contemporary British Society. Cambridge, Polity Press, Chapter 5.

Etherington, J. R. 1982 *Environmental and plant ecology*. 2nd edition. John Wiley and Sons, Chichester.

Herbert, S. Territoriality and the Police, The Professional Geographer, [on-line] 49, 1, 86–94, http://pluto.bids.ac.uk/JournalsOnline/jol_page/20JOL-1.891334892.830800

Maugh, T. H. 1984. *Acid rains' effect on people assessed*, Science 226, p. 1408–1410.

Marshall, J.S. 1997 *Cryptosporidium parvum: detection and Distribution in two Yorkshire Rivers*, unpublished PhD thesis

Ross SL 1998 Racial Differences in Residential and Job Mobility: Evidence Concerning the Spatial Mismatch Hypothesis, *Journal of Urban Economics*, [on-line] Accessed 4 April 1998

Williams M. 1991. The human use of wetlands. *Progress in Human Geography*, 15, 1, p1–22.

Brady, N. C. 1990. The nature and properties of soils. MacMillan. London. 10th edition

Young K. 1991, Shades of Green. In Jowell, Brook, and Taylor (eds.) *British Social Attitudes: the Eighth Report*, 107–130.

White, I. D et al 1992. Environmental systems – an introductory text. Chapman and Hall.

14.4 Sources and further reading

BUBL (BUlletin Board for Libraries) 1998 *Bibliography*, [on-line] BUBL Information Service, Strathclyde University, http://link.bubl.ac.uk:80/bibliography Accessed 10 January 1998.

Cross, P. and Towle, K. 1996 *A Guide to Citing Internet Sources*, [on-line] Bournemouth University, http://www.bournemouth.ac.uk/service-depts/newlis/LIS_Gen/citation/harvardsysint.html Accessed 9 January 1999.

ISO 1998 Bibliographic references to electronic documents, International Organization for Standardization, National Library of Canada, [on-line] http://www.nlc-bnc.ca/iso/tc46sc9/standard/690-2e.htm Accessed 10 January 1999.

Mills, C. 1994 Acknowledging Sources in Written Assignments, *Journal of Geography in Higher Education*, Directions, **18**, 2, 263–268.

Xia, L. and Crane, N.B. 1996 *Electronic Styles: a handbook for citing electronic information*, (2nd edn), Information Today Inc, Medford, New Jersey.

Where are we? Wordsearch 2

You are looking for 13 countries, 14 cities, 5 rivers and 3 seas. Answers on p.265.

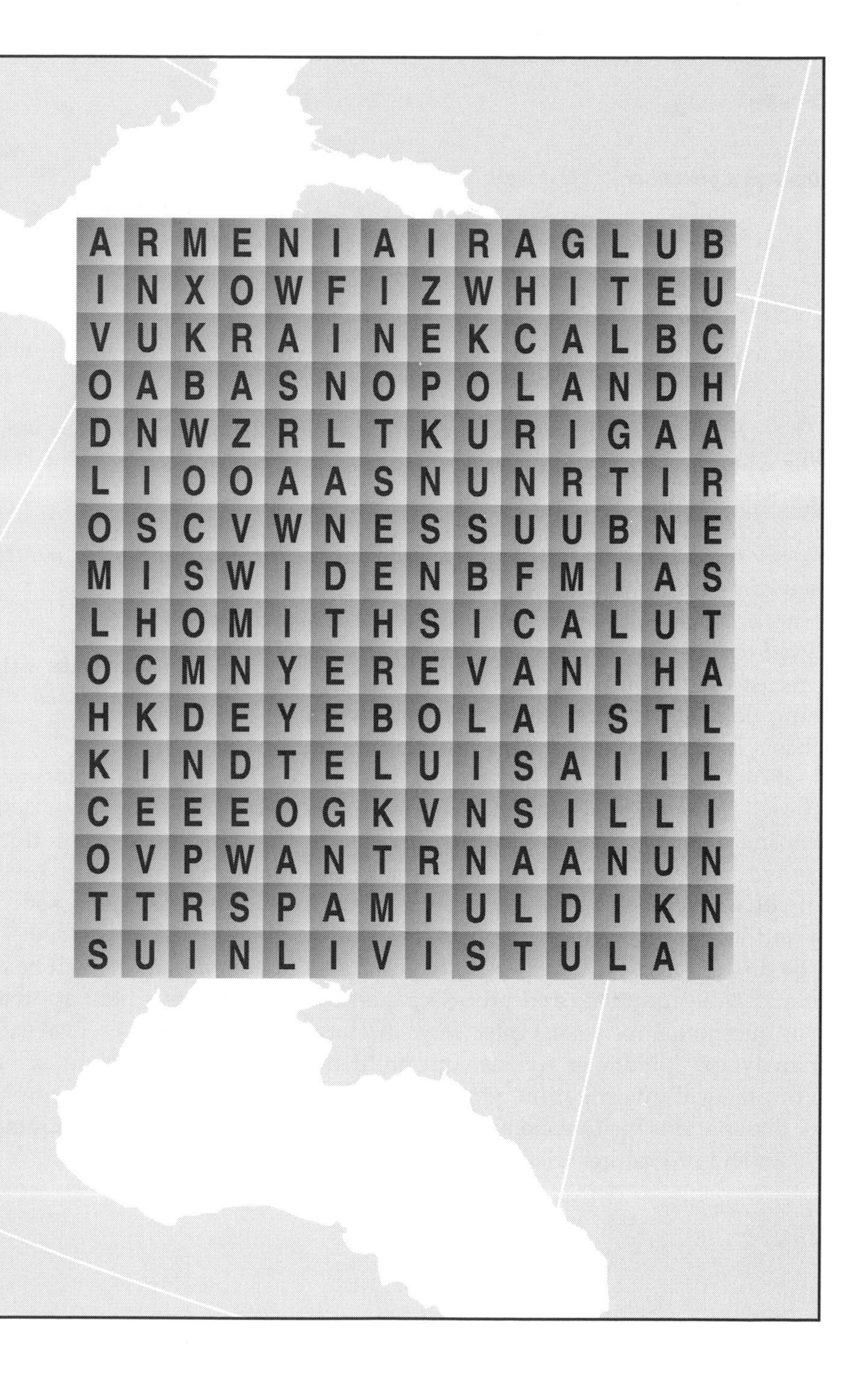

15 WRITING PRACTICAL REPORTS

'I submitted a virtual report last week.'

'Oh, OK, thanks, I'll give it a virtual mark.'

This chapter is for human geoggers too. The test of a good report is the following.

✔ The reader should be able to repeat the work without reference to additional sources.

✔ The reader should understand the significance of the outcomes of the work in the wider geographical context.

✔ It is short and to the point.

Skills involved in report writing include clear thinking, analysis, synthesis, written communication and persuasive writing.

Support arguments with evidence

Some practicals are truly *investigative*, where data is analysed to find 'an answer'. You might use census data to forecast migration patterns for the next 20 years, or housing demand in different geographical regions. The emphasis is on achieving and evaluating an answer. This type of practical is possible because you have also done 'developmental' practicals. *Developmental* practicals help you understand the *research process*, by evaluating where and how each step works. In this kind of investigation the report will place less emphasis on the final result than on the steps involved. It will involve statements about assumptions, validity of data, techniques, sources of error and bias, and alternative approaches evaluated and used or rejected. There may be discussion of the choice of statistical test, a forecasting model will be taken apart and its assumptions and processes questioned, alternative field approaches such as questionnaires versus interviews discussed. This kind of practical calls on your analytical thinking as well as your problem-solving skills. Depending on the type of practical investigation, your report will change its focus and emphasis. Many dissertations include elements of developmental practicals, questioning the processes and procedures used.

15.1 The format

The normal expectation is that the writing is brief, direct, and follows the general format outlined in Figure 15:1.

Chapters/Sections	Contents
Title	Short and precise title.
Author details	Author's name and contact address (e-mail).
Abstract	Short, precise summary.
Introduction	Background. Why geographers (and you) are doing this.
Methodology	Computer, field and/or laboratory procedures, sampling methods, site details.
Results	A summary of the findings using tables, graphs and figures.
Discussion and Evaluation	Interpret the results using statistics and modelling as needed. Consider the accuracy and representativeness of the results and their interpretation. Discuss their significance in both this experiment and their wider geographical relevance.
Conclusion	A brief summary of the outcomes. You might describe what could be done next to continue the research.
References	Vital, include methodological and geographical references.

Figure: 15.1 Outline for a report

Abstract or summary

You might write this in the style of an Executive Summary (see p.165). Explain the geographical context, hypotheses tested, methodology, main findings and interpretation in brief sentences.

Introduction

Describe why you carried out this piece of work and give a brief indication of where it fits with the rest of geographic knowledge. The hypothesis being tested should be stated precisely. Refer to other experiments and research on this topic, which might involve a mini (1–5 paragraph) literature review.

Methodology

The technical details describing the research approach, computer programs, field or laboratory procedures.

Site details A short site description and reasons for choice of site will set the scene. Use maps, grid references, sketches and photographs so the reader can locate and picture the location.

Sampling methods Outline the sampling and analytical scheme. Include maps, flow charts and equipment diagrams. This is the point to discuss potential errors, calibration issues, and the influence that the selected sampling procedure might have on the results.

Standard and non-standard procedures Many computer programs, statistical and laboratory procedures are standard. They must be cited but do not need to be explained in detail. Statements like 'Questionnaire responses were coded and entered for analysis using SPSS', or 'The standard procedure for water hardness by titration was followed (Geography Laboratory Manual Test No 65)', or 'The program was written in Grass and imported to ARCINFO for testing', will do fine. You do not need to tell the readers about SPSS, how you tested the water, or give details of GRASS or ARCINFO. **BUT** if you make up a new method, adapt a standard method or adopt a non-standard approach it must be explained in enough detail for the next researcher to evaluate and copy your technique. A full technical or laboratory protocol might be put in an appendix. Explain, also, why it was necessary to invent a new method, what was the problem with previous approaches? NB Science students have a habit of correctly including laboratory procedures, chemical lists, apparatus details and sample size information, BUT omit details of the subsequent qualitative, graphical or statistical analysis, which loses marks!

Results

Summarise the results in tables, graphs, flow diagrams, and maps to show the main findings. Label and order the diagrams so the sequence flows to match the logic of the discussion and evaluation sections. Consider putting the raw data in an appendix, or submit on disk, don't clutter this section with it.

Discussion and evaluation

This section will cross-reference to the results section with the emphasis on explaining relationships, lack of relationships and patterns in the results. Consider how the results fit with previous geographical experience. Refer back to the introductory and literature sections to place these results in a wider context.

Caution

There is usually more than one explanation for most things. All measurements are prone to error and inaccuracy, if you have not worked out what might be a problem with the data, talk to people to develop some ideas. No self-respecting examiner will ever believe that you have a perfectly accurate data set (unless you happen to be a God, in which case, my apologies for inferring your fallibility). Unless you are absolutely certain, it might be wise to use a cautious phrase like 'It would therefore appear, on the basis of this limited data, that ...', or 'Initially the conclusion that ... appears to be justified, but further investigations are required to add evidence to this preliminary statement'.

Conclusion

Resist at all costs the temptation to explain all the Wonders of the Known World from two measurements of pH in a mixture of mashed leaf, or a simulation of the 'Changing Population of the village of Pigsmightflybury 1998–1999'. Ensure your conclusions are justified by the data, not what you hoped to find. This is the place to suggest what research should be done to continue the investigations.

References

A report is a piece of academic work. Reference your sources and quotations fully (see chapter 14).

Appendices

Some information is too detailed or tangential for the main sections. It would distract the reader from the exciting storyline, but may be included in appendices. Appendices are the home for questionnaires, details of laboratory protocols, example calculations, programs, and additional data, maps and graphs. Appendices are not 'dumping grounds' for all raw data and field notes. Limit appendices as far as possible. For most laboratory reports they are totally unnecessary.

Find out what is wanted when

15.2 Laboratory notebooks

Many physical geography practicals are assessed through laboratory reports or notebooks, which are written during the practical and submitted as the laboratory class ends. This approach encourages the adoption of professional laboratory practice, noting experimental problems as they happen and making interpretations at the time. Getting good marks probably, almost certainly, indubitably, means reading handouts and doing the recommended pre-practical reading, so that you can interpret the results as you obtain them. A practical class is short and this approach seems tough, but it develops your skill in describing and analysing results at speed, describing what you actually see, rather than a fuzzy vision partially recalled three weeks later.

15.3 Where the marks go!!

Most students hope most marks are awarded for all the time spent sweating over a steamy PC, talking to billions of shoppers and persuading them to complete questionnaires, trying to log into remote computers, digging trenches across moraines and mixing sewage sludge in acid. TRAGICALLY, most academics see these efforts as worthy of some, minor, reward. A tutor has produced the handouts and briefing, explained what to do and how to go about it. The practical session will develop your skills and experience in handling data, samples or spades, and be worthy of maybe 50 per cent of the marks at the most generous. What your tutor really wants to know is:

- Have you understood what the results mean?
- Do you know why a geographer would want to be doing this in the first place?
- Do you understand the flaws in the method and the accuracy of the results?
- Do you know what the intelligent research geographer would do next?

Hence most marks go for the Discussion, Evaluation and Conclusions section of the report. So leave time to write these up, *and* redraft them so the writing is good, *and* add the references. Many reports lose marks because they stop at the Results stage.

This next sentence concluded a student report: 'It can therefore be concluded that a fuller investigation is required, increasing both the sample size and its spatial extent.' This conclusion was on the right lines and showed the author appreciated that sampling was an issue BUT it is possible for the author to add value with a conclusion that is more information rich. Something like: 'The restricted data set has limited the inferences that may be drawn. Follow-up research should incorporate a broader cross-section of respondents. If finance were available, extending the research to contrast small towns (Bridgewater, Colchester, Ledbury, and Thirsk) with major industrial and commercial cities (Bristol, Glasgow, Leeds, and Manchester) would increase confidence in the conclusions.' This is longer but no more difficult to write. The original author had discussed future experimental work informally, earlier in the project, but 'was shy of adding these sorts of ideas to the report'. Shy is nice in social situations but doesn't get marks. Forget 'shy' when reporting.

Use your departmental practical assessment sheets as a guide to self-assess and revise your reports, or use Figure 15.2. A 60/30/10 mark split is indicated here, what counts in your department?

15.4 Further reading

Flowerdew, R. and Martin, D. (eds) 1997 *Methods in Human Geography: a guide for students doing a research project*, Longman, London.

Lobban, C.S. and Schefter, M. 1992 *Successful Lab reports: a manual for science students*, Cambridge University Press, Cambridge.

Sides, C.H. 1992 *How to Write and Present Technical Information*, (2nd edn) Cambridge University Press, Cambridge.

Name .... Laboratory Report Title ....

	1	2.1	2.2	3	Fail	
Structure and Argument 60%						
Logical presentation						Discontinuous report
Topics covered in depth						Superficial report
Links to geography clearly made						No links to geographical material
Clear, succinct writing style						Rambling, repetitious report
Technical Content 30%						
Experiment appropriately described						Experiment poorly described
Graphs, tables and maps fully and correctly labelled						Incorrect/no labelling of graphs, tables and maps
Sources fully acknowledged						Incorrect/absence of referencing
Correct, consistent use of units						Units incorrect/ inconsistent
Presentation 10%						
Good graphics						Poor/no use of graphics
Good word processing						Poor/no use of word processing
Comments						

Figure: 15.2 Assessment criteria for a laboratory report

What do geographers and computers have in common?

You have to punch information into both of them.

16 REVIEWING A BOOK

Authors love criticism, just so long as it is unadulterated praise.

At some point in your student career, a tutor will explore your skill in summarising and identifying the principal points in geographical material by asking for a book review. With luck it will involve a text of only five zillion pages and a two-week deadline. This all adds to the fun (possibly), and admit it, given six weeks you would probably put off doing any preparation until two weeks in advance!

Before embarking on this exciting venture, read a couple of geography book reviews. *Progress in Physical Geography*, *Progress in Human Geography*, *Transactions of the Institute of British Geographers* and *Area* publish book reviews in most issues. Look for ideas of style and content. If you can find a copy of the book being reviewed, to compare the reviewer's comments with the original, then so much the better. Have a look at **Try This 16.1**; it gives an insight into the types of comments made by reviewers.

There is no right way to write a book review but there are some general guidelines. Book reviews are highly personal, reflecting the opinions of the reviewer, but think about the audience first. It will help you to decide where to place the emphasis of the review, and guide the formality and style of the writing. There are two general styles of review:

1. *Descriptive* – an objective summary of the contents, scope, treatment and importance of a text.
2. *Analytical or critical* – an objective appraisal of a text's contents, quality, limitations and applicability. It should discuss the text's relative merits and deficiencies and might compare it with alternative texts. This may require allocating time to browse through other material to place the book in context.

TRY THIS 16.1 – Book reviews

The following extracts are from the Book Reviews published in *Area*, **29**, 3, 1997 (page numbers indicated). To get a feel for the approach and style of review writing, make a brief comment on each of the quotations. It is hard to make objective judgements without reading the original text and the whole review, aim to distinguish style and content points. Some possible responses can be found on p.265.

1. The appearance of *Disability and the City* represents an important moment for the discipline, being the first major geographic analysis of disability in published form. (p.274).
2. However, it is the scope and scale of the empirical material presented in this volume that mark its greatest achievement. Questions of gender and race, restructuring, spatiality, consumption and politics are explored in 15 different chapters, each providing a window onto theoretical abstractions. (p.276)
3. *Glacial Geology: Ice Sheets and Landforms* is very up-to-date and well written. It covers an extremely wide range of topics and concludes with an interesting chapter on 'Interpreting Glacial Landscapes'. It is written for students and I expect that they will really like it. (p.279)
4. Because of the breadth of subject matter, certain individual sections are quite sparse. Also, I feel that in places there is a strange combination of new and old ideas presented side by side, even though they may be contradictory, e.g. there is much discussion of lodgement tills and melt-out tills but no suggestion of the debate over their distinguishing criterion and large-scale existence. (p.279)
5. It is important therefore that any introductory text concisely defines its subject matter at the outset. Nowhere, however, do we gain a clear indication of what environmental science *is* (or is *not* for that matter). (p.280)
6. The book sets out to provide a manual of landslide recognition for the non-specialist and it achieves this objective admirably. Its focus on European features might mean that some of the most spectacular world examples are not included, but it also means that there are landslides which I am sure have not previously appeared in the English literature. (p.281)
7. It has taken three years for this selection of papers to see the light of day. ... One might question the purpose of having such delayed, if thoroughly refereed, proceedings, when many of the results have been overtaken by events elsewhere, not least by the contributing authors. (p.282)
8. In the time between the San Diego meeting and the appearance of the book, other advanced texts showing the forefront of research with a better coverage have appeared, for example *Lancaster's Geomorphology of Desert Dunes* (1995) and Abrahams and Parsons' *Geomorphology of Desert Environments* (1993). (p.283)
9. These chapters are rather more variable in quality and some deserve scant attention in the opinion of the reviewer. Is it really of any cartographic value, for example, to use sketch maps produced by 124 undergraduates to postulate as to gender differences in map reading abilities (Kumler and Butterfield, Chapter 10)? How, one wonders, is the fact that apparently 32 per cent of males and 14 per cent of females put north at the top of their maps, going to add to our knowledge of map design? (p.284)
10. The writers ought to be congratulated on producing such a clear and informative text covering a broad range of issues within the field of people and the environment. It is rare for an edited volume to have the coherency contained in this text. (p.285)

To get a good perspective on a text, set aside time for reading well in advance. Leave time for your brain to develop opinions. The SQ3R technique (see p.49) is certainly valuable here, especially if you have a short time limit.

A *descriptive review* might include a combination of some of the following elements:

- an outline of the contents of the book;
- a summary of the authors aims for the book and the intended audience;
- an evaluation of the material included and comments;
- quotations or references to new ideas to illustrate the review;
- a brief summary of the author's qualifications and reference to his/her other texts;
- cite the geography texts that this book will complement or replace, in order to place this text in its academic context;
- a summary of any significant areas omitted.

Including long quotations is not a good idea unless they really illustrate a point. A reference list is required if you refer to other texts in the review.

A *critical book review* gives both information about, and expresses an opinion of, a book. It should include a statement of what the author has tried to do, evaluate how well the author has succeeded, and present independent evidence to support the evaluation. This type of review is considerably more time-consuming than a descriptive review. While reading, note passages that illustrate the book's purpose and style. Remember to balance the strengths with the weaknesses of the book and also consider how the author's ideas, opinions and judgements fit with our present knowledge of the subject. Be sure that where you are critical, this is fair comment given the author's stated aims for the text. Reading the preface and introduction should give you a clear idea of the author's objectives.

A critical review might include:

- a description of the author's purpose for writing and qualifications;
- the historical background of the work;
- the main strengths and weaknesses of the book;
- a description of the genre which the work belongs, it academic context;
- a commentary on the significance of the text for its intended audience.

Top Tips

- Read the book!
- Make notes about principal themes and conclusions, look at the book again.
- Think about the content and decide on a theme for the review, look at the book again.
- Draft the outline to support the theme of the review and check that nothing vital is missed, look at the book again.
- Draft the review, look at the book again.
- Edit and revise the final version.

You must include a full reference (see chapter 14) so a reader can locate the original text. You may also include the price and perhaps compare prices of competitive texts. Hard and softback prices can be found through booksellers or the www.

You may want to comment on the style of the writing and the ease with which you think the intended audience will understand the contents. If it is well written then say so, 'this text is clearly written with case studies and examples illustrating key points' or 'although intended as an undergraduate text its style is turgid, the average undergraduate will find the detail difficult to absorb'. People read reviews to find out which books to read, and they like to know whether the material is written in an accessible manner. The real skill in reviewing involves giving yourself enough time to absorb the content of the text and then let your brain make the connections to other pieces of reading so that you offer valid links, complements and criticisms.

A Book Review Cannot Be Completed In One Draft On The Night Before A Tutorial!

16.1 References and Further Reading

Berry, B.J.L. 1993 Canons of Reviewing Revisited, *Urban Geography*, **15**, 1, 1–3.
Hay, I. 1995 Writing a Review, *Journal of Geography in Higher Education,* Directions, **19**, 3, 357–363.

Discuss sources of error

17 ABSTRACTS AND EXECUTIVE SUMMARIES

I got the Abstract sorted, ... so that just leaves the dissertation.

Abstracts and Executive Summaries (ES) inform readers about the contents of documents. While both approaches summarise longer documents, they have different formats and serve different purposes. Writing an abstract or an ES enhances your skills in reading, identifying key points and issues, structuring points in a logical sequence and writing concisely.

Most geography journal articles include an abstract that summarises the contents. Abstracts appear in bibliographic databases to notify researchers of an article's content. Executive summaries are normally found at the start of reports and plans, particularly with business documents. An ES aims to describe the essential points within a document, usually in 1–2 pages. Depending on the context, the style may be more dynamic and less formal than an abstract.

17.1 Abstracts

Look at some well-written abstracts before writing one. *Transactions of the Institute of British Geographers, Environment and Planning, Progress in Physical Geography,* and *Progress in Human Geography* have examples and look at **Try This 17.1**. An abstract should be a short, accurate, objective summary, there is no room for interpretation or criticism. Abstracts should do the following:

- ☺ let the reader select documents for a particular research problem;
- ☺ substitute, in a limited way, for the original document when accessing the original is impossible;
- ☺ access, in a limited way through translations, research papers in other languages.

Most departments expect an abstract with a dissertation, but not with tutorial essays. If you are asked to prepare an abstract of a paper or book, it might be useful to check off these points:

- ✔ Give the citation in full.
- ✔ Lay out the principal arguments following the order in the full text.
- ✔ Emphasise the important points; highlight new information, omit well-known material.
- ✔ Be as brief, but as complete, as possible.
- ✔ Avoid repetition and ambiguity. Use short sentences and relevant technical terms.
- ✔ Include the author's principal interpretations and conclusions but do not add your own commentary. This is not a 'critical' essay.

Aim for about 80–150 words. The first draft will probably be too long, and need editing.

TRY THIS 17.1 – Abstracts

Next time you read a journal article, read the paper first and make notes without looking at the abstract first. Then compare your notes with the abstract. Are there significant differences between them? Think about how you can use an abstract as a summary. (Remember, reading abstracts is not a substitute for reading the whole article.)

17.2 Executive summaries

An effective executive summary (ES) is a very much shortened version of a document and the style is generally less literary than an abstract. The format often involves bullet points or numbered sections. The general rule on length is one side only, on the basis that really, really, really busy people will not read more. An ES can be part of an organisation's promotional material, in which case an upbeat, clear style with lots of impact is advantageous.

An ES written as part of a student exercise would normally be short, summarising a report or essay in one side. Longer ESs can be found on government web sites. These may have a PR aim, for example to acquaint the reader with government policy in an accessible manner, rather than expecting a reader to tackle a draft Act of Parliament or Congress. See the ES for the Clinton administration's policy on Welfare Reform at http://www.acf.dhhs.gov/news/welfare/regexec.htm (Accessed 10 January 1999) as an example.

The essential element when writing an ES is to eliminate all extraneous material. Do not include examples, analogous material, witticisms, pictures, diagrams, figures, appendices, or be repetitious or repetitious or repetitious. An ES will do the following:

- ✔ be brief;
- ✔ be direct;
- ✔ include all main issues;
- ✔ indicate impacts, pros and cons;
- ✔ place stress on results and conclusions;
- ✔ include recommendations with costs and timescales if germane.

Chapter headings and subheadings may present a starting scheme for bullet points. Look at the discussion and conclusion sections for the main points the author is making. You can find examples of ESs through **Try This 17.2**.

Geography students, set the task of generating an ES to accompany a report, thought it looked remarkably like a good outline plan, albeit with the emphasis on results and recommendations. Writing an ES can positively assist in structuring essays.

On p.132 the use of abbreviations was suggested as a way of speeding up writing, dealing with longer words, or with phrases repeated regularly. How do you view the use of ES rather than executive summary in this chapter? Should this style be adopted? Note that when the phrase first appears, (ES) occurs afterwards to indicate that this abbreviation will be used thereafter.

Graphs and diagrams help explanations, especially in essays

TRY THIS 17.2 – Executive summaries

Have a look at some of these on-line sites and note the different styles of ES, or search for your own sites. Executive+summary will start a search, add *+government* or *+health* to refine your search area. As with all www-related exercises some sites may be defunct and aim to work when the system is not too slow.

ESs are available, without a subscription, for *The Journal of Public Policy and Marketing Online*, at http://www.ama.org/pubs/jppm/index.asp, and *The Journal of Marketing Online*, at http://www.ama.org/pubs/jm/index.asp. Accessed 10 January 1999 [Click on 'Latest Issue' icon and then any paper title. Is there a difference here between an ES and an abstract? Is this a case of Marketing Journals being upbeat with an ES rather than an abstract?]

EPA 1997 *Benefits and Costs of the Clean Air Act*, [on-line], United States Environmental Protection Agency, Washington DC, http://www.epa.gov/oar/sect812/copy.html Accessed 9 January 1999 [The US EPA statement on the Benefits and Costs of the Clean Air Act. This document has a 2-page abstract and a 10-page ES, compare them.]

Goldie-Scot, D 1997 Banking and finance in South Africa, [on-line] http://www.live.co.uk/ftsa.htm Accessed 10 January 1999

Kitchin, R., Shirlow, P. and Shuttleworth, I 1998 On the Margins: Disabled People's Experience of Employment in Donegal, West Ireland, *Disability and Society*, 13, 5, 785–806 [on-line] http://www.qub.ac.uk/saru/dondisexec.html Accessed 10 January 1999

Lafferty, W. and Meadowcroft, J. 1997 *Implementing Sustainable Development in High Consumption Societies: A Comparative Assessment of National Strategies and Initiatives,* [on-line] http://www.shef.ac.uk/uni/academic/N-Q/pol/susdev.htm Accessed 10 January 1999

The Scottish Office 1998 *Circular On Housing And Neighbour Problems Dealing with Nuisance and Anti-Social Behaviour* [on-line] http://www.scotland.gov.uk/library/documents-w2/hanps-00htm?1;lib-d.htm Accessed 10 January 1999

The Scottish Office 1998 *Drinking Water Quality in Scotland 1997* [on-line] http://www.scotland.gov.uk/library/documents-w2/dwq97-02.htm Accessed 10 January 1999

UNDCP 1997 *World Drug Report*, [on-line], United Nations International Drug Control Programme, Vienna, http://www.undcp.org/undcp/wdr/wdr.htm Accessed 10 January 1999

Wells, J.W. 1997 *Department of Pesticide Regulation Strategic Plan* [on-line], California Environmental Protection Agency, Sacramento, California, http://www.cdpr.ca.gov/docs/planning/execsum.htm Accessed 9 January 1999

18 DISSERTATIONS

The First Law of Dissertations: *Anifink tht kin goo rongg, wyll.*

A dissertation is significant part of many geography degree programmes. Each department has its own timing, style, and expectations of length and monitoring procedures. Most departments run briefing sessions so everyone is aware of the rules. Missing such briefings is a BAD idea; one of your better ideas will be to re-read the briefing material every six weeks or so to remind yourself of milestones and guidelines.

This chapter does not in any way attempt to replace or pre-empt departmental guidelines and advice. It does aim to answer some of the questions asked by students in the first two years of a degree when a dissertation is sometimes viewed as a kind of academic Everest, to be assaulted without aid of crampons or oxygen, and to say something about timescales for effective planning. More detailed information will be found in Parsons and Knight (1995), Burkill and Burley (1996), Flowerdew and Martin (1997) and Lindsay (1997). Skills addressed during dissertation research include autonomous working, setting and meeting personal targets in research, professional report production and problem solving.

18.1 A DISSERTATION IS NOT AN ESSAY, WHAT IS THE DIFFERENCE?

Getting a dissertation together is a very different activity, requiring different skills, to assembling an essay. You look in detail at a particular topic, take stock of current geographical knowledge of an issue, and offer some further, small, contribution to the geographical discussion. It is an opportunity, given a great deal of time, to explore material, develop an idea, conduct experiments, analyse information and draw mature conclusions from the results. The results need not be mind-blowing, discovering a new continent is rare, and the world will not end if your research has a completely unexpected answer.

Essentially a dissertation is an opportunity to enquire systematically into a topic or problem that interests you, and to report the findings for the benefit of the next person to explore that material. Dissertations are rarely published, but aiming to produce a product that is a worthy of publication is appropriate.

18.2 Time management

Typically you will explore a topic by yourself, making all the decisions about when, where, how, and in what detail to work. Supervisors will advise, give clues as to what the department expects, but basically it is down to you to plan and organise. Look at the handing in date, say 1 May of final year, now work backwards:

1. Allow three weeks for 'slippage', flu, visitors, Easter, career interviews, despondency (April).
2. Writing up time – allow 3–4 weeks – OK, it should take 6–7 days but you have other things to do in March (March).
3. Analysing the data, fighting the computer systems, getting print outs (February).
4. Recovering from New Year, start of year exams, career interviews (January).
5. End of term Balls, parties, preparing for Christmas, end of term exhaustion, Christmas (December).
6. Get data from the Local Authority, Environment Agency, Census, field collection, laboratory analysis, ... allow 6–8 weeks (October and November).
7. Have an idea, check out the library, think about it, talk to supervisor, have two more ideas ... settle on a topic, allow 6–8 weeks.

A final-year project or dissertation, designed to be done between October and May, will consume chunks of time through the period. With an early October start date (meaning late October, because weeks 1 and 2 are needed for recovering from the vacation, starting new modules, catching up on friends and partying), a 1 May finish is very close. It is worth asking about planning in year 2. Can fieldwork or data retrieval be completed in advance? Can you check out potential field sites at Easter of year 2? Summer vacation fieldwork, and especially fieldwork abroad, needs advanced planning, the right equipment, clear formulation of hypotheses, and a workable, planned timetable.

What are the supporting ideas?

A dissertation should not be rushed if it is to get a high mark. Most low marks for dissertations are won by those who start very late, and those who leave the crucial thinking elements to the last week. You may have noticed useful thoughts occurring a couple of days after a discussion or meeting. You wake up thinking 'Why didn't I say ...' High marks become attached to dissertations in their third rather than their first draft. This means being organised so that there is thinking time at the end.

18.3 Choosing topics

You need an idea to explore or hypothesis to test that is geographical in nature and, ideally, one that captures your imagination. If you are not interested, you are unlikely to be motivated to give it time and thought. If you fear your future might involve panning for small change from cardboard city, consider a dissertation that is relevant to possible areas of employment, demonstrating your interest and skills to a future employer and giving you an insight into marketing, pollution control, ... A dissertation may present an opportunity to indulge a hobby, like mountain biking, or to visit someone, BUT must address a good question, as in 'the impact of mountain biking on soil erosion', or 'immigrant labour in Toronto suburbs, (because my aunt lives there)'.

Any piece of research involves exploring a topic, examining it from a number of aspects and looking for solutions, interpretations or answers to issues or problems. The most awkward part of a dissertation is sorting out the questions to address. A good research topic will address an interesting question in a way that can be answered, to reasonable extent, in the time and with the facilities that are available to you. Aim for a focused topic and avoid the 'Splodge' approach (Kennedy 1992) where too huge a topic is discussed too generally, like 'a study of Vancouver/rainfall/geo-economics'.

Questions that start How ...? To what extent ...? Which factors cause ...? will to focus thinking. Trying to come to judgements may be more difficult, as in Are Keynesian economic values more useful than ...? Topics that are both very big and largely unanswerable should be avoided, like Will geography have relevance in economic planning in the second half of the twenty-first century? and avoid those where data are impossible to obtain, as in A discussion of the landscapes of Pangea at latitudes 25–45ºN, or An evaluation of the geomorphological history of the sixteenth moon of Saturn through analysis of the chemistry of surface samples.

Think small at the start.

Spotting gaps

Throughout the degree course make notes of thoughts like: 'Why is it like that?', 'Is it really like that in the supermarket, town, glacier, desert that I know?', 'Is that really right?', 'But I thought ...' Also when lecturers say, 'but this area is not researched', or 'this was investigated by x in 1936, no developments since'. Another entry point is when an argument seems to have gone from alpha to gamma without benefit of beta. Now, it may be that the lecturer does not have time to explore beta on the way, there may be a great deal known. Alternatively, beta may be unexplored, a little black hole in search of a torch.

Having spotted an apparent 'hole' take a couple of hours in the library and on the www to see what is available. Many a tutor has sent a student to research a possible topic, to be greeted by the response that 'the topic is not on because there are no references available'. SUCCESS IS FINDING LITTLE OR NOTHING. This is what you want, a topic that is relatively unexplored so that you can say something about it. Researching for an essay you want lots of definitive documentary evidence to support arguments. For a dissertation you want to find little, or contradictory material, to support the contention that this is a topic worth exploring. If there is lots of literature, that's OK, use it as a framework, a point to leap off from, to explore and extend. Watch out for areas of the subject where the published literature goes out of date very quickly. Anything to do with government policy, such as housing, changes with a new party in power – and can change rapidly within the lifetime of an administration as policy evolves. You can investigate the current position.

Browsing is a primary dissertation research technique. Try to immerse yourself in material that is both directly related, and tangential to the topic. Wider reading adds to your perspectives, and should give an insight into alternative approaches and techniques. You cannot use them all, but you can make the examiner aware that you know they exist.

Topics

There are still lots of questions to be answered, all the lines of geographical enquiry have not been sorted out, although one or two are done to death. Topics go in cycles. There are fads that relate to whatever is taught in the two weeks before dissertation briefings, and standard chestnuts often expressed as 'I want to do something on air quality; global warming; water chemistry changing downstream; crime in the city; social housing.' These are quite positive statements compared to the student who wants to do 'something historical' or 'something to do with weather' or 'something while canoeing in Scotland'. The choice of dissertation is your responsibility. There may be a departmental list of suggested topics and past dissertations can be read in the library, but thinking of topics and deciding on a valid research hypothesis is part of your dissertation activity.

Take care if you choose to investigate a personal interest. You might do an extremely good dissertation on 'badger habitats', 'the environmental impact of the Felphersham bypass' or 'the impacts of EC legislation on the fishing industry in the 1990s'. You will get into deep water, and probably lower marks, if you bias your answer with anti-badger baiting literature only, relate your activities as 'my summer as Swampy's tunneler', or present an entirely pro-Greenpeace fishing story. Writing any of these stories would be fine in a newspaper article that seeks to put across a single viewpoint. A dissertation is an objective academic exercise and so requires objective reporting. If you feel extremely strongly about an issue, you might write an excellent or a direly unbalanced report. Think about it.

Survey-based studies

For geographers, collecting and evaluating data is the normal stock in trade. Hence most degree courses include modules on document analysis and survey techniques including questionnaire design, observation and interview techniques (Lindsay 1997, Marshall and Rossman 1995). Results are reported through quantitative and qualitative summary of the responses and statistical analysis where appropriate. Where data are required on a longer historical scale, secondary data such as the census, national or government data, national and international statistics and library archives are vital sources.

Collecting your own data has the advantage that you know where and when it was compiled, you have a feel for its accuracy and know exactly why you asked particular questions. The disadvantages include the time-consuming nature of repetitive sampling and some consideration of sample representativeness is required. The time available for data collection is short, so data are often representative of one day, week or month in one season, at a limited number of sites. This is better than no data, and a perfectly good way to proceed. Be wary of over-generalising from the individual to the global case. Avoid inferences like 'The 30 grains of sand monitored in the first order tributary of the River Swale moved 6mm in 12 weeks in 1999, therefore we can conclude that River Nile sediments have moved an average 16 miles since the birth of Christ.'

The critical issue with secondary data is access. Data always takes longer to arrive than you hope. Can you discover why it was collected originally? The purpose of the original collector, and your purpose, are almost certainly different. This is not a problem, but you need a paragraph somewhere to tell the reader you understand where the original data collection methods and collation processes impact on the results. Think about data decay issues too. Does your data really characterise the period you are interested in, or is it X years old? It is tempting to match the best secondary data available with current observations and not to address the data age gap. It may not matter or it might be really important, depending on your topic. Ask whether the arbitrary nature of sampling influences results. For practical reasons sampling is undertaken on an hourly, weekly, bi-monthly time-scale; does this introduce bias in results?

Case studies

Exploring 'Social diversity in two villages in Eriador' or 'The Recent Ankh-Morpork Flood' can be great. Case studies allow detailed examination of a specific area, time period, incident, event, cross-section or computer simulation. The aim is to understand more about how the world normally operates by examining a typical example. Alternatively, you might want to explore the exceptional case to see how the general model responds in extreme circumstances. Watch out for conditions where your 'typical study' deviates from the average.

Time series or cross-sectional studies

Time series studies let you examine the way different groups respond to the same phenomena, or to a changing situation. You could look at shopping patterns in people in each 10-year age cohort. This allows you to explore, and compare behaviour and attitudes, to differentiate patterns of shopping in the under-10s, teenagers, and older cohorts. Take care to remember the influence on different cohorts of cultural and social influences. You may wish to explore changing attitudes and conditions with individuals, thereby generating longitudinal information.

In physical geography, downstream studies of sediments or water quality, weather pattern change over time and soil catena investigations are regular topics. Ensure the background control conditions are kept as constant as possible. It should be no surprise to discover bedload chemistry changes downstream if the river passes a number of chemically distinct bedrocks, two mines and three sewage treatment works on the way.

Theory-based topics

An examination of the geographical theory related to a part of the subject is perfectly acceptable and potentially very rewarding, though may seem a little risky at first glance. What will not get high marks is a voyage into your personal opinions on the Weberian approach to social class, or a personal diatribe on what you think are the causes of all the ills of a city. Tackling a theory-based dissertation allows you to think through a series of related issues but there must be evidence to support the argument. This requires careful inspection of information. Be especially careful to locate the supporting and detracting arguments, and to come to a balanced judgement about the full range of material. Set the theory in its geographical context, which requires wide reading around alternative positions.

Before embarking on a theoretical study, talk the ideas through with your tutor, who will be enthusiastic and have good advice about the ways in which you might develop your thinking. Having read and thought further, go back and discuss the new ideas. Theoretical work needs to be bounced off people, so talk to your housemates, tutor group and other geographers including postgraduates if possible. Share your ideas and become familiar with them. Browsing and discussion are essential elements of the theoretical geographers' research methodology.

Developing or evaluating a technique

This makes a good dissertation if careful attention is paid to sample size and getting the statistics right. You might look at the accuracy of an instrument; at variations on mathematical equations; a sensitivity analysis of a model process; or at the comparative accuracy of measuring something with three or more techniques. Comparing gps (global positioning systems) instruments is topical at present. The advantage is that the problem is defined fairly tightly and, provided enough measurements are made, the precision of the results can be evaluated.

What if . . . project

This is a speculative dissertation, and requires sensible thought and valid forecasting frameworks. It must be a topic where you can gain some information, whether from data or historical expertise to build the argument, and the counter argument. 'An evaluation of the geographical consequences of re-nationalising the UK railway system in 2002' would be a speculative piece of research, embedded in an examination of the historical consequences of the last nationalisation of the railways and commenting on issues surrounding privatisation. Reporting a number of 'what if' scenario forecasts using transport and economic models will ground the speculations and arguments. This type of dissertation demonstrates synthesis skills.

Beware

There is a temptation to feel you should show one thing or another, that there is a black and white element to the issue. In most geographical situations there are large areas of both subjective and objective disagreement. Much of what we know is an approximation, we hope to make the best inferences, given the information gathering procedure available at the time. Interview-based information is subjective, relying on people telling the truth and there being time to collect enough material. Even objective data has its inaccuracies. Population geographers can supply you with data on a country's population, but they will also tell you that even where the census is well regulated, individuals avoid the count for personal reasons, and errors occur in collection and calculation procedures. The estimated historical population of England in 1777 is just that (estimated), for 977 and 477 'population facts' are debatable estimates. Knowledge is partial, go for your best interpretation on what you know now, but keep looking for and recognise areas of doubt in your arguments.

A tutor is important especially in warning against falling for trite arguments that can appear beguiling. Watch for moral arguments which are acceptable in one culture or time, but inappropriate in another culture or time; evaluating Third World development polices using the standards and attitudes of a twenty-first century northern European will not work. Arguing that something 'is the best' is also fraught with danger. There is a good case to argue that Humboldt is the greatest geographer, but that accolade, I would argue, personally, goes to either Captain Cook or Jacques Cousteau. Some less well-read persons might want to give this credit to Johnston, Haggett or ... (these two sentences could keep a tutorial group going for some time). These last few sentences are extreme, unbalanced and unsupported as they stand. Avoid unsupported arguments and strong language unless there is overwhelming evidence.

18.4 Resources

Is the data you need available at the scale you want, and in the timeframe available to you? There may be plenty of Spanish customs documents for trade in the nineteenth century, but if they are locked in a Cadiz office and you do not read

Spanish, they are no good to you. Finding out about the availability, scale and accuracy of the data that you would like to use, is a task to complete early. If the data does not exist at the right scale and in enough detail, your project may need to be rethought. Do a pilot study.

You may need hard and software. Check out the library, computer and laboratory opening hours. Do you need to book equipment in advance? Have you booked the laboratory equipment for the days after sampling? One day's idyllic rowing across a lake collecting water samples may involve 5 days' laboratory analysis. In the vacation that may mean organising accommodation and time off work.

18.5 Pilot surveys

Just about every lecturer nags every student to do a pilot survey, about 90 per cent of students do not bother and about 89 per cent get into a mess as a result. PILOT SURVEYS ARE A GOOD IDEA. They allow you to grab a few samples, or trial a questionnaire or run a mini-programme. Then you analyse the pilot dataset as if it was the entire set, right through laboratory, statistical and modelling procedures, to putting dots on graphs and numbers into equations. Life is too complicated and the environment much too complex for anyone to get everything right first time. A pilot survey shows there are other angles, other people to consult, suggests other variables to measure, that a screwdriver could be essential and just being in the field, on site, at a computer or whatever, indicates other things you could do to enhance the research. More than one student has discovered that their ideal site has been cleared for development, or been built over since their last visit. At the very least, go and look at a site before doing oodles of preparation. If you are working overseas then practise your techniques on comparable local sites before leaving. Make sure you have all the right equipment and a complete checklist of the data you need.

At the end of a pilot survey ask yourself:

- ? Is this the best approach?
- ? Can I get all the data needed to answer the hypothesis?
- ? Can the project be done in the time allowed?

If not, revise the methodology.

18.6 The format and assessment

A dissertation or thesis is a formal piece of writing. The reader expects the author to adhere to the 'rules', make the presentation consistent and tidy and observe word length rules. If possible, ensure that the same typeface is used throughout, that figure titles appear in consistent positions, that there are page numbers on ALL pages, that all the figures and tables are included, and that you double check the university guidelines and follow them, especially the WORD LENGTH. The general format is roughly along the lines given in Figure 18.1. Treat the page lengths as suggestions. There are plenty of high quality dissertations submitted in different page combinations.

Find the marking guidelines your department adopts or see Figure 18.2. There are many variations, some departments ask for an initial plan that counts, others do not. Remember the percentage distributions are only a guide, mark distributions vary considerably depending on the nature of the dissertation.

Chapter	Contents		Page Length
0	Title Page, Acknowledgements, Abstract, Table of Contents, Table of Figures.		
1	Introduction, Brief background, Research aims, Signpost thesis layout.		2–3
2	Literature Review, summary of material relevant to this research. Links to geography generally.		4–8
3	**Theoretical Thesis**	**Empirical Thesis**	
	Methodology, a description of your research approach *or* Discussion and evaluation of first Theme/Idea/Concept.	Methodology – your research process, techniques used, criticism of techniques and evaluation of their accuracy and representativeness. Site information.	2 – piece of string
4	Discussion and evaluation of Second Theme/Idea/Concept/	Results, with tabulations, graphs, and maps, as required.	Piece of string
5	Discussion and evaluation of third Theme/Idea/Concept/ Counter-themes. Synthesis of themes and alternatives.	Interpretation of results – this may include further modelling work. Evaluation of accuracy and representativity, and sensitivity analysis if relevant.	Another piece of string
6	Implications for future research. *'What I would have done if I had known at the start what I know now.'* Conclusions		1–2 1 1
Appendix	Data sets if required. Example copy of Questionnaire. Computer program. Sample of interview transcripts.		Minimise

Figure: 18.1 Typical dissertation formats

18.7 Writing and Writing Up

This 'writing up' activity rather implies that it can be done in one go, and as a one off. Developing ideas and seeing the implications of results takes time. Remember

Dissertation Assessment	Please comment using following headings as appropriate
Planning Phase *Clarity in formulation of project? Originality in formulation of project? Independent development of project?*	**10%**
Abstract	**5%**
Literature Review *Relevance of literature selected? Comprehensive? Critical comments on literature?*	**15%**
Methodology *Appropriate to topic? Successful in execution? Followed plan and adapted it appropriately?*	**20%**
Analysis and Interpretation *Planned? Appropriate? Extent to which aims were met? Consciousness of limitations?*	**25%**
Discussion and Conclusions *Logical and thought through? Sustainability of conclusions? Suggestions for future research?*	**15%**
Presentation *Quality of figures, tables, maps, and photographs? References? Appendices? Page Numbers? Appropriate Length?*	**10%**
Degree of Supervision Required *Did the student take the initiative? Any illness or personal problems?*	
Further Comments	

Figure: 18.2 Dissertation assessment criteria

to write as you go. GET SOMETHING ON PAPER EVERY WEEK. Read a couple of things and then draft some paragraphs for the literature review. You can add to it when you read the next paper. Writing is part of the research activity, you read a bit, write a bit, think a bit; and the combination of these three activities tells you what you might do next. The dissertation can also exemplify your time management skills. Typical dissertation research patterns are shown in Figure 18:3, the optimist's pattern is sadly misleading.

You cannot hope to have read all the past literature and to research all aspects of a topic. Much of dissertation management is about drawing a line in the sand, stopping reading, stopping investigating and starting writing. You are aiming to report on what you have read and discovered. Your opinions are based on stated material. If you took another year over your dissertation, you would still be reporting on partial information. Your examiners are looking to see that you have tackled a reasonable topic in a relevant manner and drawn sensible inferences and conclusions from the findings. They do not expect you to model the processes of global warming at the planetary scale, account for the social development of all Latin American communities or explain why buses come in threes. Don't get overwhelmed by reading, keep it balanced. Use the checklist in **Try This 4.6** (see p.43) as a guide and remember to balance 30 minutes of www, or on-line bibliographic searching, with at least a couple of hours of reading.

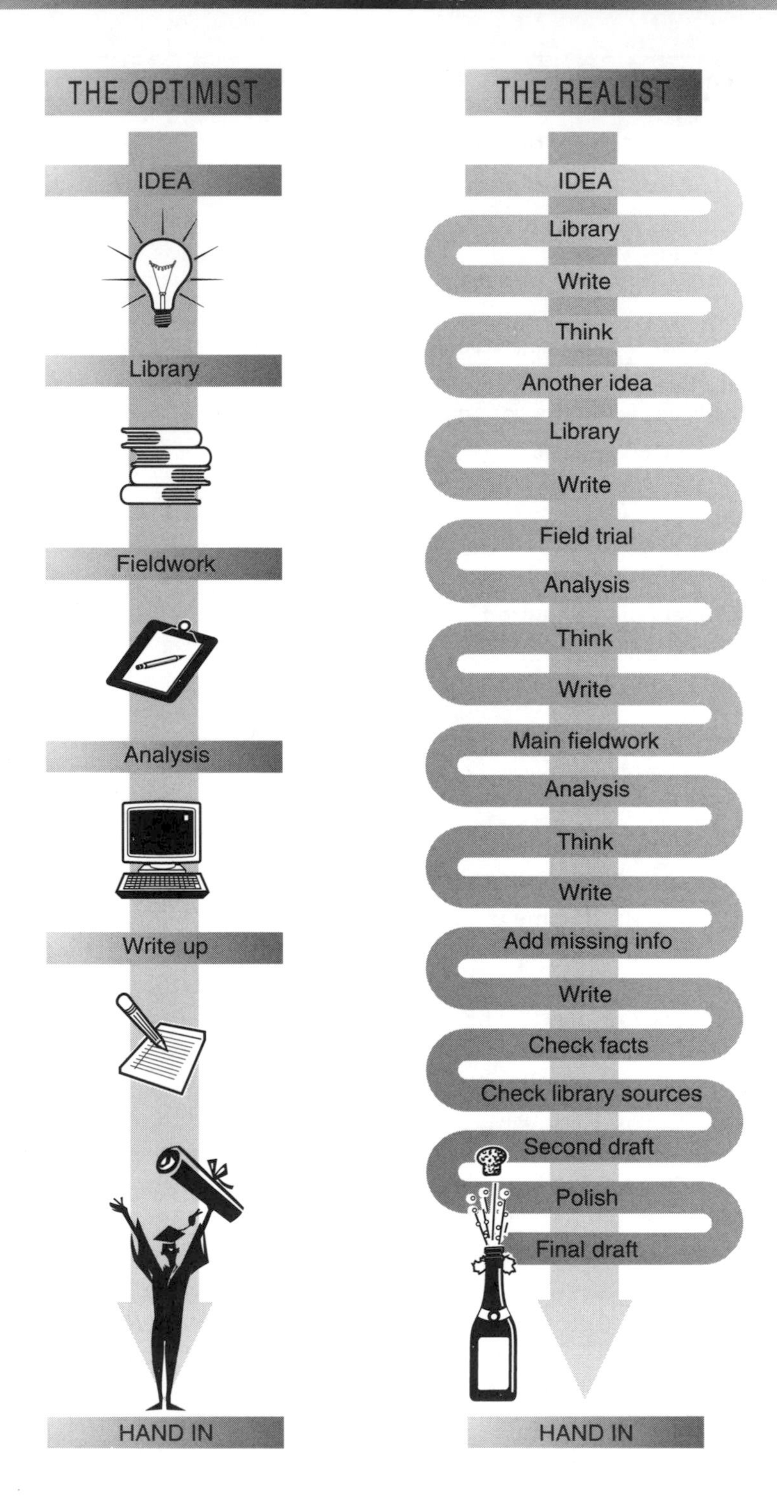

Figure: 18.3 The process of dissertation research

- **Start writing**, the draft will not be right the first or second time. Ask tutors how often they rewrite before sending a piece to a publisher. 'Lots' is the only answer worth believing.
- SPELL CHECK EVERYTHING
- **References** matter, check that your references are properly quoted. It is miles easier to do this as you go along. Put them at the end of the document from the start. Getting to the last three days and then realising you have kept no record of references is like getting to within 6 inches of the summit of Everest and having to go back to base camp for flag and camera.
- **Proof-read** what you have written not what you think you wrote. Check for titles and keys on graphs and maps, scale bars, units with equations and data, and that the title page, abstract, acknowledgements and contents page are in the right format for your department.
- Proof-reading is not easy, get a flat mate to read through for grammar, spelling and general understanding. If your friend understands your 'Environmental Impact Study of Camels in Klatch' then so, probably, will your tutor. You can repay the favour by checking out your mate's dissertation on the 'The Humour of Baldrick'.

  If you can leave a little time between writing and proofing, you spot more errors.
- **Abstract** Write this last.

18.8 Safety issues

And if you break your leg on fieldwork, don't come running to me.

Dissertation research should be good fun, but accidents can happen, so be aware of your responsibilities and safety procedures. Safety is as important for human geographers in town, as for speleologists down a pothole. Each university and department has its own safety policy, and you will be given advice, but remember you are responsible for your own actions. Most safety is about common sense and taking sensible precautions. Fieldwork for the dissertation is likely to be a solo activity, but do not work alone. Offer 'a day out' opportunity to friends and relations, two heads are better than one, and the company will keep you cheerful. It is your responsibility to do the following:

- ✔ Take and keep medication or special foodstuffs with you at all times.
- ✔ Provide yourself with clothing and boots or wellingtons that are appropriate for your site conditions.

- ✔ Check weather forecasts.
- ✔ Leave a plan and map showing where you intend to go with someone who will alert the authorities if you have not returned by a set time. Remember to check in with them when you get back, the police and mountain rescue get very unhappy searching for people tucked up in the local pub reliving the delights of the day.
- ✔ Consider taking a mobile phone to alert your contact of a change of plan or a problem. In remote areas, taking a first aid kit, map, torch, compass, spare food and a flask with a hot drink is sensible.
- ✔ Take no unnecessary risks.

Generally, fieldwork safety advice is aimed at physical geographers heading for uninhabited areas of Scotland, Iceland or Tunisia. Human geographers also need to take care. Working in towns and cities can be as hazardous. Take care on unfamiliar roads, and especially in countries where drivers are on the other side of the road. Be courteous at all times with the public. People are often curious. Have a clear, non-confrontational explanation for your activities and explain why you have chosen the particular area. Think about your clothing and ensure it is appropriate for the locality. If you are working abroad, and not fluent with the language, ask someone to write a short statement explaining what you are doing and who you are, preferably on university notepaper and with a contact telephone number. If you are heading for the mountains or glaciers you may be asked to do a mountain safety course or see Barry and Jepson (1988) , and Koester (1991).

If you are working in a developing country look at Robson and Willis (1994). It was written for postgraduates but applies equally to undergraduate visits whether for human or physical geography .

18.9 References and further reading

Dissertation research:

Bell, J. 1993 *Doing your Research Project*, Open University Press, Buckingham.

Burkill, S. and Burley, J. 1996 Getting Started on a Geography Dissertation, *Journal of Geography in Higher Education*, Directions, **20**, 3, 431–7.

Flowerdew, R. and Martin, D. (eds) 1997 *Methods in Human Geography: a guide for students doing a research project*, Longman, London.

Kennedy, B.A. 1992 First Catch your Hare ... Research Designs for Individual Projects, in Rogers, A., Viles, H. and Goudie, A. (eds), 1992 *The Student's Companion to Geography*, Blackwell, Oxford, 128–34.

Lewis, S. and Mills, C. 1995 Field Notebooks: a student's guide, *Journal of Geography in Higher Education*, Directions, **19**, 1, 111–14.

Lindsay, J.M. 1997 *Techniques in Human Geography*, Routledge, London.

Marshall, C. and Rossman, G.B. 1995 *Designing Qualitative Research*, (2nd edn) Sage Publications, Thousand Oaks, London.

Parsons, T. and Knight P. 1995 *How To Do your Dissertation in Geography and Related Disciplines*, Chapman and Hall, London.

Robson, E. and Willis, K. 1994 *Postgraduate Fieldwork in Developing Areas: a rough guide*, Monograph 8, Developing Areas Research Group, Institute of British Geographers, London.

Safety:

Barry, J. and Jepson, T. 1988 *Safety on Mountains*, British Mountaineering Council, Manchester.

Koester, R.J. 1991 *Wilderness and Rural Life Support Guidelines*, dbS Productions, Charlottesville, VA.

St John's Ambulance and British Red Cross 1992 *First Aid Manual*, (6th edn) Dorling Kindersley, London.

Geo-quick crossword 2

Answers on p.267

Across

1 Head count (6)
5 Supporter (4)
8 Erupts on Sicily (4)
9 The widest view (8)
10 No fractions here (8)
11 Slat or batten (4)
12 Caravan expedition (6)
14 Forceful apple catcher (6)
16 Just a lake (4)
18 Moisture (8)
20 Man made, far attic (anag) (8)
21 The Irish House (4)
22 Hillside vicar (4)
23 The last! holiday place (6)

22 Hillside vicar (4)
23 The last! holiday place (6)

Down

2 Tea is on (anag) (7)
3 Affirm, plight (5)
4 A great road? (12)
5 Draught (3, 4)
6 Eccentric orbiter (5)
7 Measurer of slopes (12)
13 Resource (7)
15 Isolated spot (7)
17 Misjudgement (5)
19 Extinct birds (5)

19 REVISION SKILLS

'I know I knew that once. . . . What was it?'

'I finish five modules this week, have three term papers to hand in on Friday, and the first exam is on Monday. So when do I revise?' Well, the answer is that you don't revise in this style. Revision, in the last-minute 'cramming' style, is really a school concept. It is part of a 'Get the Facts ➜ Learn them ➜ Regurgitate in Examination' process. It is characteristic of surface learning. In 'real life', you don't revise, you have an accumulating body of knowledge and apply it continuously. University is a transition phase, but all your deep reading and learning activities should mean revision as cramming becomes a small proportion of your activities.

Re-reading and reviewing material needs to be an ongoing process, built into your weekly timetable because it improves the amount of detail that is recalled. So normal learning activities should reap rewards at exam time because they are also 'revision' activities that reinforce your learning, and you have been doing them all term. The following will help:

If you know the technical geographical term, use it

- ✔ Get your brain in gear before a lecture by reading last week's notes.
- ✔ Re-read all discussion activities.
- ✔ Use SQ3R (see p.49) in reading and making connections to other modules.
- ✔ Think actively around issues (see chapter 7).
- ✔ Check that you have good arguments (see chapter 8), spot gaps in the evidence and use that information to determine what you read, rather than just taking the next item on the reading list.
- ✔ Actively ask questions as you research (see p.69).

Timetable 'revision slots' to continue these learning processes.

Tight deadlines are often a feature of university life. Most tutors will be blissfully unaware, and emphatically unsympathetic when you have five essays due on the last day of term or semester and examinations two days later. Sorting out schedules is your problem, not theirs, and an opportunity to exercise your time management skills.

19.1 Revision aims

Exams are designed to test your understanding of inter-related geographical material from coursework, personal research and reading. You will to have overview information on a large number of topics and detailed examples and case studies to back it up.

Understanding, relevance, analytical ability and expression are listed by Meredeen (1988) as the Holy Grail of examiners, keen to donate marks. Reflect on where you can demonstrate these attributes to your nice, kind examiner:

1. UNDERSTANDING Have you shown you understood the question? Keep answers focused.
2. RELEVANCE There are no marks for irrelevant inclusion of material no matter how geographical. Stick to the issues and points the question raises. Don't let yourself be side-tracked.
3. ANALYTICAL ABILITY Aim for a well reasoned, organised answer. Show that you understand the meaning of the question and can argue your way through points in a logical manner.
4. EXPRESSION Clear and concise writing, that makes your ideas and arguments transparent to the examiner, and maps and diagrams help enormously. The length of an answer is no guide to its effectiveness or relevance. A well-structured short answer will get better marks than a long answer padded out with irrelevant details.

19.2 Get organised, time management again

If from the start, you take 30 minutes each week for each module, to review and reflect, to think around issues, and decide what to read next, you will also be revising. It is an ongoing process, and you will be well ahead of the majority! As exams get closer, sort out a work and learning timetable, say, for the five weeks before exams. Put every essay, report and presentation deadline on it, add the exam times, lectures, tutorials and all your other commitments, and have a little panic. Then decide that panicking wastes time and sort out a plan to do 6 essays and revise 12 topics and include social and sporting activities to relax, and block the time in 2-hour slots, with breaks to cook and eat. Then organise group revision, discussion and sharing opportunities.

Top Tips

- Start earlier than you think you need to!
- Put revision time into your weekly plan, use it to think.
- Keep a revision record and attempt to allocate equal time to each paper.
- Speaking an idea aloud, or writing it down, lodges information in the brain more securely than reading.

Revising is not something you have to do alone. Think with friends and colleagues. Group revision assists:

☺ by generating comments on your ideas, adding the perceptions of others to your brain bank;

☺ because it is easier for two or three brains to disentangle theories explaining decentralisation or drumlins;

☺ because it's more fun (less depressing?);

☺ by making you feel less anxious about the exam, more self-confident;

☺ because the challenge of explaining something to a group will help you understand the points and remember details more clearly;

☺ by showing where you need to do extra study and where you are already confidently fluent with the material.

Don't be put off by someone claiming to know all because they have read something you have not. No one reads everything. If it is so good, ask for an explanation.

19.3 What do you do now?

What were your thoughts as you left your most recent exams? Ignoring the obvious 'Where's the bar?', jot down a few thoughts and look at **Try This 19.1**.

TRY THIS 19.1 – Post-examination reflections

Write down your thoughts about your last examinations, and look at the list below. These are random, post-examination thoughts of rather jaundiced first-year geographers. Tick any points you empathise with. Make a plan.

- ☺ No practice at timed essays since A-level, I'd forgotten how after eighteen months.
- ☺ Not enough time to write.
- ☺ A lot of the lecture detail seemed unnecessary.
- ☺ Too many specific places and dates to remember.
- ☺ Lack of revision.
- ☺ Missed the point of the last essay.
- ☺ Didn't know what was relevant.
- ☺ Was thinking about going to the bar after, most of the time.
- ☺ Needed more direction towards the questions.
- ☺ You couldn't learn the whole course; I picked the bits that were missed.
- ☺ Had loads of facts in the lectures, but nothing to apply it to.
- ☺ Mostly relied on the A-level stuff and made something up.
- ☺ I had all this stuff about Calcutta and he asks about Brazil.
- ☺ I think the detail in the lectures was confusing and the big handout added another layer of detail that wasn't in the lecture. I didn't really get started on it because there was too much to tackle.

19.4 Good revision practice

The rubrics for papers are usually displayed somewhere in a department, possibly on the examination noticeboard. Find out about the different styles of questions, how many questions you have to answer and what each question is worth. Plan your revision activities to match the pattern of the paper.

Revise actively

Sitting in a big armchair in a warm room reading old notes or a book is almost guaranteed to send you to sleep! TRY some of these ideas.

- ✔ Make summary notes, ideas maps, lists of main points and summaries of cases or examples.
- ✔ Sort out the general principles and learn them.
- ✔ Look for links, ask questions like:

- Where does this fit into this course? – essay? – other modules?
- How important is it? – a critical idea? – detailed example? – extra example to support the case?
- Is this the main idea? – an irrelevance?

✔ Write outline answers to essays, then apply the criteria of Understanding, Relevance, Analysis and Expression. Ask yourself 'Does this essay work?', 'Where can it be improved?'

✔ Practise writing a full essay in the right time. Sunday morning is a good time! Put all the books away, set the alarm clock and do a past paper in the right time. DON'T PANIC, the next time will be better. Examination writing skills get rusty, they need oiling before the first examination or your first answers will suffer.

✔ Remember you need a break of five minutes every hour or so. Plan exercise into your revision time, swimming, walking or going to the gym; the oxygen revitalises the brain cells (allegedly).

✔ Apply the ideas in the Active Reading SQ3R technique (see p.49) to your revision. Use it to condense notes to important points.

✔ Aim to review all your course notes a week before the exams! Then you will panic less, have time to be ill without getting behind, and feel more relaxed for having had time to look again at odd points and practise a couple of timed essays. (OK, OK, I said AIM!)

✔ Staying up all night to study at the last minute is not one of your best ideas in life.

Reviewing the first four weeks of a module should, ideally!! (ho-hum), be completed by weeks 6–8, so that the ideas accumulate in your mind. The rest of the module will progress better because you understand the background and there is time to give attention to topics in the last sessions of the module. But then, who is ideal?

Outline essays

Outline answers are an efficient revision alternative to writing a full essay. Write the Introduction and Conclusion paragraphs in full, NO CHEATING. For the central section, do one sentence summaries for each paragraph, with references. Add diagrams, maps and equations in full, practise makes them memorable. Then look carefully at the structure and balance of the essay, consider where more examples are needed and whether the argument is in the most logical order.

Keywords

Examiners tend to use a number of keywords to start or finish questions. Words like Discuss, Evaluate, Assess, Compare and Illustrate. They all require slightly different approaches in the answer. Use the **Try This 19.2** game to replace keywords and revise essay plans.

TRY THIS 19.2 – Replacing keywords in examination questions

Take a question, any question that has been set in an examination or tutorial in your department. Do an outline structure for your answer. Then replace the keyword with one or more of Assess, Compare, Criticise, Explore, Illustrate, Justify, List, Outline, Trace, Verify. Each word changes the emphasis of the answer, and the style and presentation of evidence.

Here is a list of keywords used in essay questions and 'possible' meanings, (adapted from Rowntree, 1988, Lillis, 1997).

Analyse	Describe, examine and criticise all aspects of the question, and do it in detail.
Argue	Make the case using evidence, for and against a point of view.
Assess	Weigh up and make a judgement about the extent to which the conditions in the statement are fulfilled.
Comment	Express an opinion on, not necessarily a long one, BUT often used by examiners when they mean Describe, Analyse or Assess (cover your back).
Compare	Examine the similarities and differences between two or more ideas, theories, objects, processes, etc.
Contrast	Point out the differences between ... could add some similarities.
Criticise	Discuss the supporting and opposing arguments for, make a judgements about ...; show where errors arise in ... use examples.
Define	Give the precise meaning of ..., or Show clearly the outlines of ...
Describe	Give a detailed account of ...
Discuss	Argue the case for and against the proposition, a detailed answer. Try to develop a definite conclusion or point of view. See Comment!
Evaluate	Appraise, again with supporting and opposing arguments to give a balanced view of ... Look to find the value of ...
Explain	Give a clear, intelligible explanation of ... Needs a detailed but precise answer.
Identify	Pick out the important features of ... and explain why you picked this selection.
Illustrate	Make your points with examples, or expand on an idea with examples. Generally one detailed example and a number of briefly described, relevant supporting examples make the better argument.
Interpret	Using you own experience explain what is meant by ...
Justify	Give reasons why ... Show why this is the case ... Need to Argue the case.
List	Make a list of (usually means a short/brief response), notes or bullets may be OK.

Outline	Give a general summary or description showing how elements interrelate.
Prove	Present the evidence that clearly makes an unarguable case.
Relate	Describe or tell a story, see Explain, Compare and Contrast.
Show	Reveal in a logical sequence why ... see Explain.
State	Explain in plain language and in detail the main points.
Summarise	Make a brief statement of the main points, ignore excess detail.
Trace	Explain stage by stage ... A logical sequence answer.
Verify	Show the statement to be true, the expectation is that you will provide the justification to confirm the statement.

The quiz approach

Devising revision quizzes can be effective. You do not need to dream up multiple answers, but could play around with a format as in Figure 19.1. This is a short factual quiz on ozone, based on a journal article by Drake (1995), but tragically the answers in the second column are out of order. Can you sort them out? The answers are at the end of the chapter.

1. When were worries about ozone first expressed?	*a. Spring*
2. What units are ozone measured in?	*b.* $X + O_3 \rightarrow XO + O_2$ $XO + O \rightarrow X + O_2$ *net* $O_3 + O \rightarrow O_2 + O_2$ *where X may be H, OH, NO, Cl or Br*
3. What is the Montreal protocol?	*c. Early 1970s*
4. What are the major anthropogenic sources of ozone destroying compounds?	*d. Dobson units*
5. In which season does the ozone level at Halley Bay, Antarctica begin to fall?	*e. Farman, J.C., Gardiner, B.G. and Shanklin, J.D. 1985 Large losses of total ozone in Antarctica reveal seasonal* ClO_x/NO_x *interaction,* Nature, *315, 207–210*
6. Which paper first reported the decreases in ozone concentration in Antarctica?	*f. Nitrogen-based fertilisers, chloride oxides particularly from CFCs, exhaust gases from supersonic aircraft*
7. How is ozone destroyed in the the stratosphere?	*g. The 1987 international treaty to limit the production of ozone-depleting substances*

Figure 19:1 Quiz questions on ozone for a journal article by Drake (1995).

The questions in Figure 19.1 cover the facts, and are good for MCQ revision. Now consider the style of these questions.

- What initiated the early concerns about depleted ozone levels in the Antarctic?
- What are the strengths and weaknesses of the Montreal Protocol?
- Why does the polar vortex form over the Antarctic but not over the Arctic?
- What effect do polar stratospheric clouds have on ozone destruction?
- When and why do ozone levels fall in the Antarctic?
- What are the implications of continued reduction in ozone levels over the next 30 years?

These questions demand the same initial, factual knowledge as those in Figure 19.1, but a more reasoned and extended response. They are useful for revising short answer questions and essay paragraphs.

To get into the swing have a go at **Try This 19.3** and **Try This 19.4**. The first asks for quiz questions based on your notes from lectures and additional reading, and the second uses the same technique with a journal article.

TRY THIS 19.3 – Quiz questions from notes

Pick a set of notes and create a short quiz, 10 questions, in each of the two styles. First, short factual questions, then some more extended questions. Write the questions on one side of the page and answers on the reverse, so you can use them (without cheating too much) for revision.

TRY THIS 19.4 – Quiz questions from a journal article

Pick an article from any reading list and, rather than making notes, devise a set of 5 short and 5 extended questions that explore the topic. Again, write the questions on one side of the page and answers on the reverse, for revision purposes.

Revision is an opportunity to think about and continue to look for geographical explanations and insights. Revision is therefore a creative process, with ideas gelling and developing as you re-read and reconsider. Note making is an integral part of revision. Some general questions running in your head will encourage a questioning, active approach to revision. Have a go at **Try This 19.5** to explore this approach further.

TRY THIS 19.5 – Generic questions for revising geography!

Compile a list of questions that could be used in compiling a revision quiz. Some suggestions are on p.267, but have a go at your own before looking.

19.5 References and Further Reading

Drake, F. 1995 Stratospheric Ozone Depletion – an Overview of the Scientific Debate, *Progress in Physical Geography*, **19**, 1, 1–17.

Meredeen, S. 1988 *Study for Survival and Success*, Chapman, London.

Rowntree, D. 1988 *Learn How To Study: a guide for students of all ages*, (3rd edn) Warner Books, London.

Answers to Figure 19.1

1c, 2d, 3g, 4f, 5a, 6e, 7b.

Geograms 4

Try these geographical anagrams. Answers on p.268.

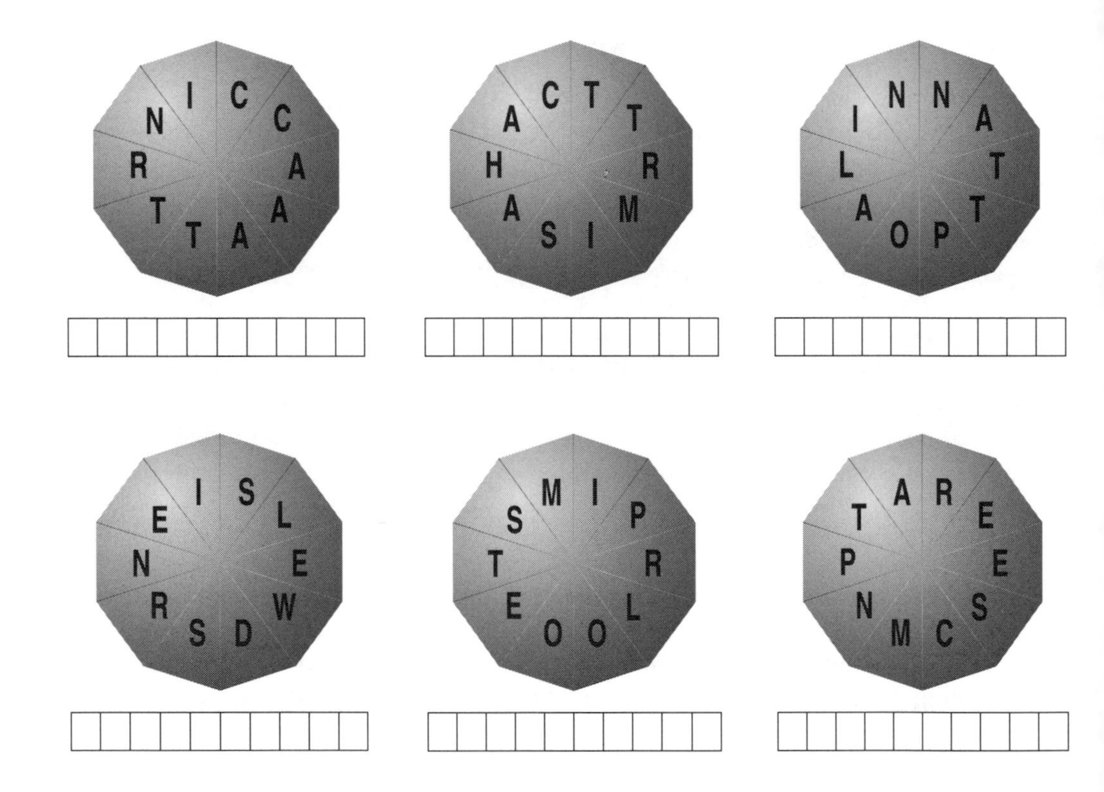

20 EXAMINATIONS

Describe the geomorphological history of Europe from the Silurian to the present day. Outline the nature of process response following the evolution of humans, and contrast with the position in Australia during the same period. Be brief, concise, and specific, using maps and 3D block models where relevant. Time Limit 1 hour.

Relax, you did all that reflection and reviewing so the examinations will be most agreeable. Check, and double-check, the examination timetable and room locations, they can change. Know where you are going, plan to be there 20 minutes early to find your seat or block number, visit the loo and relax. Check the student handbook so you know what to do if you are delayed. Make sure you have a pen, spare pens, pencils and calculator if required. Make sure you read the information at the top of the exam paper and take heed of anything the invigilator has to say. If you think there is a problem with the paper tell an invigilator at once, so the geography staff can be consulted.

Find out in advance how the paper is structured and use the time in proportion. For example, on a two-hour paper that has 6 short questions and 1 essay there will probably be 50 per cent for the essay and 50 per cent split equally between the 6 short questions. One hour for the essay and 9 minutes for each short question, leaving 6 minutes to read the paper and plan the answers. Answer the number of questions required, no more and preferably no less. Leave time to do justice to each question, and don't leave your potentially best answer until the end. Equally, don't spend so much time on it that you have to skimp the remaining essays. Make sure there are no questions overleaf.

Should you be seized with anxiety and your brain freezes over, use the 'free association', brainstorming approach. Write out the question and then look at each word in turn scribbling down the first words that occur to you, anything ... authors' names, examples, and related words. This should generate calm and facts, and you can plan from the spider diagram you have generated.

20.1 Examination essays

The minimalist advice here is re-read the advice on argument (Chapter 8), revision (Chapter 19) and writing essays (Chapter 12), and write fast. Essays allow you to develop lines of argument, draw in diverse ideas and demonstrate your skills in argument, analysis, synthesis, evaluation and written communication. Remember to keep the geography content high, use evidence to support your arguments wherever you can and cite supporting references.

Top Tips

- If all questions look impossible, chose the one where you have the most examples to quote, or the longest question. Long questions usually give more clues to plan the answer. 'Discuss the role of demand and supply mechanisms as they operate through the goods market, the resource market and the informal market', is a question with loads of clues and parts to answer, whereas 'Discuss the role of demand and supply in the geography of marketing' is essentially the same question. It could be answered with the same information, but will be a minefield if you do not impose structure and facts.
- Plan your answer even if short of time. Underline keywords in the question, like Discuss, or Compare and contrast, and note the spatial and temporal scales the answer should cover (see Fig. 20.1). Don't restrict examples to one country if the question asks about global or international issues. Do a quick list or spider diagram of the main points, and note ideas for the introduction and conclusion. Then rank the points to get a batting order for the sections.
- On a three-question paper plan Questions 2 and 3 before writing the answer to question 2. Your brain can run in background mode on ideas for Question 3 as you write the second answer.
- Watch the time. Leave a couple of minutes at the end of each answer to check through, amend spelling, add extra points, references and tidy diagrams and maps with titles, scales and units!

Some of the points examiners search for may be deduced from Figure 20.2, which lists comments from a random selection of examiners. They concentrate on the

Channel River Corridor Floodplain Catchment		1:1 1:50 1:1250 1:10,000 1:50,000 1:1,000,000	Leeds West Yorkshire Yorkshire and Humberside United Kingdom European Union	mm m 100m km 1000km	cm^2 m^2 hectare km^2
Second Minute Day Year Decade Century	Household Ward Town County Region Nation	By-Laws National Law International Law	Parish Council Local Authority Regional Council National Legislature International Grouping	Genetic Diversity Species diversity Ecological diversity	

Figure: 20.1 Geographical scales

Does not address the question 25%
Swarming with factual errors 35%
Fine answer to a question about National Parks, pity the Question was about Country Parks 38%
Very weak effort – no attempt to explain examples or definitions 42%
Somewhat confused but some relevant points 42%
Good background but little analysis on the question set 44%
A scrappy answer with some good points 45%
Started fine then got repetitive 49%
Decent attempt, some examples but not focused on the question 50%
Not sufficient explanation/evaluation, you list some ideas at the end but these needed developing to raise the mark 52% (2 sides only)
Reasonable answer but misses out the crucial element of ... 54%
Has clearly done some reading but fails to write down the basics 54%
Reasonable effort as far as it goes, but does not define/explain technical terms – no examples 55%
Accurate but generally descriptive, never really got to the 'evaluate' section 55%
Needs to learn about paragraphs, then fill them with organised content 55%
Reasonable effort – covers many relevant points, structure adequate but you need to organise points in shorter paragraphs 57%
Too much description – not enough argument 58%
Good – but needs to be more focused to key points 58%
Good discussion but needs to emphasise geography more 60%
Good introduction and well argued throughout, although with no evidence of reading. Good use of examples but unfortunately only from class material 64%
Has done some reading and thinking 65%
Quite good – a general answer, omits to define land-use but lots of non-lecture examples 64%
Excellent 70%
Outstanding 85%

Figure: 20.2 Comments on examination essays

lower end of the marking scale and from it you may gather that examiners cannot give marks unless you tell them what you know: explain terms, answer all parts of the question, use lots of examples, remember you're a ~~Womble~~ geographer. Everything needs setting in a geographical context. If you do not include geographical cases and examples, relating theory to the real world, you are not going to hit the high marks. Organise your points, one per paragraph, in a structure that flows logically. Set the scene in your first paragraph and signpost the layout of the answer. Be precise rather than woolly, for example rather than saying 'Early on . . .', give the date; or for the general 'In UK towns we see . . .' use examples like 'In Winchester, Reading and Shrewsbury we can see . . .'. Try to keep an exciting, interesting point for the final paragraph. If you are going to cross things out, do it tidily.

Take-away or take home examinations

Essays written without the stress of the examination room can seem a doddle, but still need revision and preparation. Get the notes together in advance, make sure you have found the library resources you require before the exam paper is published; at that point people are trampled underfoot in the race to the library. Think about possible essay topics to get your brain in gear. Do not leave it all to the last minute (Hay, 1996a).

20.2 Short answer questions

Short answer questions search for evidence of understanding through factual, knowledge-based answers and the ability to reason and draw inferences. For short answers a reasoned, paragraph answer is required to questions like 'Briefly describe humanism', 'Outline the difference between a geography of women and a feminist geography' or 'What was Mackinder's contribution to geographical theory?' Questions like 'Outline four characteristics of the new cultural geography' or 'Give four explanations for the decentralisation of retailing' require four, fact rich answers, and can be answered as a set of points. Take care to answer the question that is set – 'Define and demonstrate the importance of the following elements in social survey: a stratified random sample and participant observation'. This is a really helpful question if you answer all 4 parts! See examples on p.84 *et seq*.

2.3 MCQs

MCQs (multiple choice questions) test a wide range of topics in a short time. They may be used for revision, in a module test where the marks do not count, or as a part of module assessment where the marks matter. A class test checks what you have understood, and should indicate where more research and revision is required. In a final assessment watch the rules. With on-line assessment, once the answer is typed in and sent, *it cannot be changed*.

Look carefully at the instructions on MCQ papers. The instructions will remind you of the rules, such as:

There is/is not negative marking. (With negative marking you lose marks for getting it wrong). *One or more answers may be correct, select all the correct answers*. (This is how you can get 100 marks on a paper with 60 questions).

General advice says to shoot through the paper answering the all questions you can do easily, and then go back to tackle the rest (but general advice does not suit everyone). Questions come in a range of types:

The 'Trivial Pursuit' factual style

These test recall of facts, and understanding of theories, usually a small proportion of the questions:

CPRE stands for: A) Centre for the Promotion of Regional Excellence,

B) Capital Protected Rate of Exchange,

C) Council for the Protection of Rural England,

D) Centre Parcs Rafting Exercise.

Darcy's Law is usually expressed as: A) Q = CIA B) Q=KIA

C) Q = K(h/l)A D) Q= CID

In this example there are two correct answers, B and C, and both should be indicated for full marks.

Reasoning and application style

Reasoning from previous knowledge gives rise to questions like:
Which of the following sequences correctly ranks air pollution emissions in the UK in 1994, (lowest emission first)?

A) Nitrogen oxides, Sulphur dioxide, Black smoke, Carbon monoxide

B) Black smoke, Nitrogen oxides, Carbon monoxide, Sulphur dioxide

C) Black smoke, Nitrogen oxides, Sulphur dioxide, Carbon monoxide

D) Black smoke, Sulphur dioxide, Carbon monoxide, Nitrogen oxides

Some questions give a paragraph of information and possible responses. You apply theories or knowledge to choose the right response or combination of responses. Some tests combine information from more than one module, as here where a second-year soil analysis paper requires information from first and second-year statistics modules:

A soil survey of 25 fields yielded measurements of NO_3, NH_4, P, K, particle size, OM, and bulk density, crop yield (tonnes per acre). The fields were planted with 4 crops, (8 of barley, 8 of broccoli, 2 of sugar beet and 7 of broad beans).
1. Which statistical tests might be used to analyse the data set comparing production of all 4 crops? Tick all the correct answers: A) Mann Whitney U, B) Student's t-test, C) Correlation, D) Multiple regression, E) One-way Analysis of Variance, F) Histograms, G) Chi-square H) Nearest Neighbour analysis, I) Kruksal-Wallis test, J) Two-way Analysis of Variance.

2. Which statistical tests should not be used to compare broccoli and sugar beet yields? Tick all those not correct: A) Mann Whitney U, B) Correlation, C) One-way Analysis of Variance, D) Histograms, E) Chi-square, F) Nearest Neighbour analysis, G) Kruksal-Wallis test, H) Two-way Analysis of Variance, I) Student's t-test, J) Multiple regression

Data response style

A combination of maps, diagrams, data matrices and written material are presented, and a series of questions explore possible geographical interpretations.

Appropriate revision techniques for MCQ papers include deriving quiz questions. Revisit **Try This 19.3** and **Try This 19.4**. Generating questions forces you to concentrate on details.

20.4 Laboratory examinations

Some laboratory courses have associated examinations. Taking laboratory notebooks or manuals into such exams may be allowed, if so, do not forget yours. Questions tend to be based on applying practical laboratory experience to geographical issues. Revising with past papers is always a good practice, or make up your own questions. Try variations on the following:

? *If soils experiment X was rerun with soil samples from Exmoor / Tunisia / Spitsbergen how would you expect the results to change?* Essentially a 'what will happen if the samples come from different environments' question. They are designed to see if you understand controlling principles.

? *The flume tests were re-run with flows of XXXX $m.s^{-1}$. How would you anticipate the sediment erosion and deposition patterns to alter from those observed in the experiments?* Another 'what if we change the parameters' question that asks you to speculate logically about the outcome.

? *After Experiment 12 was completed the XXXX meter was found to have an error of 15%. Explain how this information affects the interpretation and how it can be accounted for in the calculations.* An 'how can you cope with instrument and operator error' question.

? *Explain where errors can arise in … (water sampling, pollen counting) … and the impact they have on data analysis and interpretation.* A general question about where errors arise in your research.

? *Describe the safety implications involved in XXXX, and explain how they can be minimised.*

? *Explain how and why instruments to measure XXXX should be calibrated before field use.*

? *Outline the field sampling and laboratory tests you would use to explore XXXX. How accurate would you expect your results to be?* A question that asks you to look creatively at how you would apply skills acquired in laboratory classes to answer geographical hypotheses, and to consider the relative adequacy of each technique.

20.5 Oral examinations

Vivas

Vivas exist as part of some modules in some degree schemes, and in some universities as part of the final examination process. The thought of a viva has been known to spook candidates. Don't panic! Vivas are an opportunity to talk about a topic or subject in detail, and is a skill that should be more widely practised, since in the workplace, you are much more likely to 'explain an idea or project' than to write about it.

The finalists *Viva Voce* are generally part of the degree classification process. After all the assessments and examinations, when all the marks are sorted out some people are 'borderline specialists'. People so close to the boundary that another chance is given to cross a threshold. A viva may also be given to candidates who have had special 'circumstances', such as severe illness, during their studies. The viva is usually run by the external examiner, a delightful professor or senior lecturer from another geography department. DO NOT PANIC. In all the departments I know vivas are used to raise candidates. You have what you have on paper, things can only GET BETTER. So if your name is on a viva list that is good news, sigh with relief, remember all that revision you did before and check out some answers to the questions below. Get a good night's sleep, avoid unusual stimulants (always sound advice) and wear something tidy. Examiners tend to appear in suits but are interested in your brain not your wardrobe. Celebrate later.

Typical viva topics

Dissertations are a frequent opening topic. You are the expert, did all the research and wrote it up. So the **Top Tip** is to glance through the dissertation, it is on a disk somewhere. Rehearse answers to questions like :

? What are the two main strengths of your dissertation research?

? Please will you outline any weaknesses you feel the research has?

? Can you explain why you adopted your research strategy?

? What do you feel were the main geographical issues you addressed?

? Since you started on this topic I see you did a module on ... How might you have adapted your research having done this module?

? Can you talk a bit about where your results relate to geography generally?

? Obviously in a student dissertation there is limited time for data acquisition, how might you have expanded the data collection process?

? Would you like to expand on the literature review/results section/ interpretation?

? Could you talk about sources of errors in the data?

Try to keep answers to the point and keep up the technical content. Any examiner recognises flannel. Waffle cannot reduce a mark, but you are trying to raise it. External examiners have supervised, marked and moderated thousands of dissertations during their career. They know every dissertation has strengths and weaknesses, they are impressed by people who have done the research and realise there were other things to do, other ways to tackle the issue, other techniques, more data to collect ... so tell him/her that you know too.

Use your 'geography speak' skills. Asked about the 'urban village concept' you could say: 'It was an idea that started in America in the 1960s, and a number of people wrote about it. Then urban villages declined as economic pressures changed. People moved out to the suburbs. More recently there has been re-investment in the city centre leading to an increase in village style communities within city centres. In the UK the idea of urban villages is beginning to catch on with planners'. It is a general answer but in the 2.2 class. You could say: 'Gans described the urban village concept in 1962. He looked at the way Italian-American immigrants to Boston lived an essentially rural lifestyle within the city, preserving traditional social structures and in some cases their language. The break-up of these core communities followed the transfer of industry and populations to the periphery of the city. However, in the last ten years the regeneration of inner city areas, gentrification of property, an increase in service sector jobs in the city centre and a desire to live in the centre, close to work and leisure opportunities have revitalised city centres, and urban villages are re-emerging albeit in a different form. In the UK in 1997 the Urban Villages Forum published a description of urban villages and explained why they can be useful in planning British cities.' A longer and more geographical answer, the references are there, Boston is an example and there are geography terms like core, periphery, social structures, and regeneration. This answer has the evidence and should get a First. So think a bit and answer with evidence, and examples, and references, just as in an essay.

After that, I think you cannot do much more preparation. The examiner can ask about any paper and question, about the degree as a whole and geography in general. If you missed a paper or question, or had a nightmare with a particular paper, and if you are still awake, think about some answers to the questions BUT the odds are they will not be mentioned.

After every exam, aim to review your answers as soon as you can. What did you learn from the exam process? University exams are a little different to school ones, a spot of reflection on your revision and exam technique might be useful before next time. Questions that might occur to you include:

? Had I done enough revision? The answer is almost always no, so ask '*Where could I have squeezed in a little more revision?*'

? Was there enough detail and evidence to support the answer?

? Would different revision activities be useful next time?

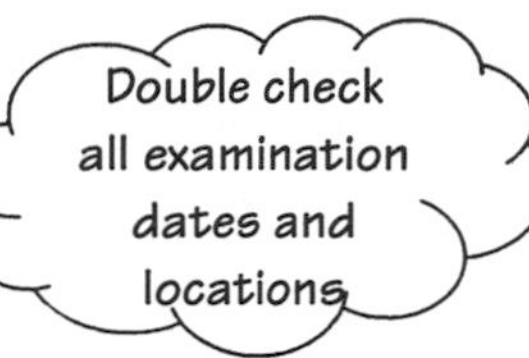

20.6 REFERENCES AND FURTHER READING

Davies, D. 1986 *Maximising Examination Performance: a psychological approach*, Kogan Page, London.

Hay, I. 1996a Examinations I: preparing for an exam, *Journal of Geography in Higher Education*, Directions, **20**, 1, 137–142.

Hay, I. 1996b Examinations II: undertaking an exam, *Journal of Geography in Higher Education*, Directions, **20**, 2, 259–264.

Geographical links

Add 2 letters in the middle squares to complete the 5-letter geography related words to left and right. When complete, an 8-letter geographical term can be read. Answer on p.269.

S	H	I			A	C	H
F	U	N			R	T	H
R	A	D			I	O	N
C	A	N			I	E	N

21 FIELDCLASS

'A geo-political boundary is an imaginary line between two nations, separating the imaginary rights of one, from the imaginary rights of the other.'

Fieldtrips are one of the great bonuses of geography degrees. While the occasional student might view a fieldclass as a fantastic holiday with mates, the staff tend to have a learning agenda that is well planned and costly. Geography departments donate a considerable proportion of their budgets to fund fieldwork because they are convinced it does the following:

☺ develops observational skills;

☺ gives a realistic insight into research activities;

☺ allows hands on, practical experience of field techniques;

☺ encourages individuals to take responsibility for research decisions;

☺ develops group and inter-personal skills;

☺ encourages students to talk through issues with staff, through small group research work;

☺ allows students to experience alternative and sometimes unfamiliar environments.

Skills acquired through fieldwork may include organisation, independent and team work, measurement, description, safety, use of maps and statistics, and presentation through posters, field notebook, reports and discussion. Add these and other skills from your own experience to your CV.

21.1 When is it?

The dates for residential classes are usually known a year in advance. Find out the dates and plan accordingly. If you have a first year fieldtrip, the dates should be advertised in pre-course literature together with information on costs. Do not book skiing holidays, footie trips or family silver wedding celebrations during this period, because fieldtrips are usually a whole, or part of a module, so the marks count.

21.2 Time management

Generally these are easy modules to manage because the staff organise the programme. Expect a programme that is something like: breakfast at 8, leave at 9, exciting field activities 10–5, dinner at 6, analysis, discussion and report from 7.15, hand in written report at 9 the next day. Field classes prove you can get up and be ready to work at 9 every day for a week, despite having enjoyed the company of colleagues for much of the night. Parents are prone to wonder why you need three days' sleep on return from field class. Tell them something comforting about sad, sadistic, but fit, lecturers that made you walk nine miles each days, AND collect data, AND use computers, AND write it all up before bed AND your room mate snored – or something.

21.3 Preparation

So you thought a glacier was a bloke who fixed windows?

Practical

Six weeks before the field-class check out your passport, injections, holes in wellies, and arrange to borrow rucksacks, wet weather gear, clip boards, and pens and pencils that keep writing no matter how hard it rains. Ask people who went last year what you really need to take BUT just because last year there was full sun on Denbigh Moor does not mean the temperature can be guaranteed to rise above arctic survival bag point this year. Unreasonably, geography staff want to view the landscape from the top of the hill, and the hill is rarely equipped with a road or a path. Consider borrowing boots. By week zero you should be ready.

Academic

Where are you going? It makes sense to look at a map and find out a bit about the place whether it is Bangor or Bangalore. Some departments have pre-departure exercises and briefings, others rather assume that as an intelligent 18+ person you will know that it may snow in the Mediterranean in January and that you need drachmas to buy drinks in Greece. Get to know what to expect. Ask last year's group. Reminiscing about field class over a few pints is normal for geographers. If completing a field notebook is part of the assessment have a look at Lewis and Mills (1995). If video activities are involved see Lee and Stuart (1997).

21.4 Safety

Someone will alert you to fieldwork safety issues. Most departments have safety codes and some will ask you to sign to confirm that you are aware of safety issues and your responsibilities. Generally you will be expected to do the following:

- ✔ co-operate with the staff about safety, and behaviour in the field;
- ✔ act in such a way that neither you, nor anyone else in the group, are put at risk;
- ✔ not wander off or do something hazardous;
- ✔ make staff aware in advance if you have a medical condition, allergies, asthma or diabetes for example;
- ✔ know what to do if something goes wrong, like becoming detached from the group, missing the bus or becoming disoriented if the weather deteriorates and visibility is lost;
- ✔ have appropriate clothing.

The staff will have completed risk assessment documentation, planned for problems and gained permission for site access. They will probably have been on training courses on first aid and safety. They will check the weather and adjust the fieldclass timetable and activities accordingly.

In rough terrain, field paths, hills, anywhere out of town, field boots should be worn. For river work, or on peatbogs, wellingtons with soles that grip are most useful. Aim to have lots of thin layers, T-shirt, shirt, sweater and fleece under a cagoule is ideal. Layers can be removed, or added, during the day. A woolly hat and gloves are essential for fieldwork in the UK from September to May, and waterproof trousers and a bright cagoule with hood to go over an anorak or fleece are recommended. Good field gear is expensive; can you borrow clothes for a short fieldclass? Everyone on fieldwork should be sensible, polite, considerate of others, and take especial care not to damage anything from hotels to field walls, or to leave gates open.

Safety is mostly applied common sense. As a general rule of thumb 'if you wouldn't do it with Grandma or a baby on a Sunday walk', then probably it is a bad idea; like standing in the middle of roads to sketch buildings, trespassing on private property, entering caves, derelict buildings, or any water that is more than wellington boot deep. Stick to public footpaths and rights of way. Wear hard hats in quarries, under cliffs and on steep slopes where rock or debris falls are a hazard. Do not climb on cliffs or rock faces. Take care on foreshores, rocky beaches and keep an eye on the tide. Do not light fires and take especial care in woodland and on moorland to ensure cigarettes are extinguished. Soil and water can carry bacteria, viruses and other pathogens that cause disease. Use appropriate sampling methods, gloves when taking water samples for example and ensure that you wash your hands before eating.

Insurance arrangements vary between institutions. You may be asked to insure yourself or to contribute towards the costs of insuring equipment while in the field. Make sure you know what is required.

Take care and have fun, most fieldclasses are very well planned and free of incidents.

And while away, be environmentally friendly. Use recycled toilet paper.

21.5 Virtual fieldclasses

Traditional fieldclasses are held in real field locations, complete with wind and weather. Expense and timetable issues are leading some departments to develop virtual field trips, but you do not have to wait for your department to take you. Virtual fieldwork can be accessed via the www and some sites provide useful research material and jolly pictures! You can access the Virtual University site at: http://www.uwsp.edu//acaddept/geog/projects/virtdept/vfthome.html (Accessed 10 January 1999). If this address no longer works, do a web search using a combination of Virtual+Fieldtrip+Geography+Texas.

If a nice sunny island field visit will cheer you up, have a look at the Hawaii trips, sun, sea, and the film locations for Jurassic Park and some geography. For a more harrowing visit, the Bosnia trip will update you on political and associated issues in the Balkans, ethnic cleansing, the involvement of the USA and much more. With all web visits be aware that authors may have a particular point of view to promote. Don't put your critical faculties on one side while on virtual fieldtrips. With all web sites check when it was last updated. Fieldtrips to the San Andreas Fault or Antarctica are relatively timeless. Sessions which deal with issues like Bosnia or Iran become dated very quickly, what has happened since? However, a Web site two years old is probably more up to date than a textbook of the same age.

A list of virtual fieldclass sites is held at http://www.geog.le.ac.uk/cti/virt.html Accessed 10 January 1999.

Virtual fieldclasses can also be frustrating, you may not have the capacity to download pictures unless you have a high capacity pc, and staff may have denied access to parts of the site, so their students cannot see the answers before they do the trip! As with all web activities, try when the lines are quiet.

Fieldwork, virtual or real is a short experience. Staff aim to give a 'taste' of research activities. The speed of a fieldclass may mean your technical achievements are limited. At the end of a real or virtual field visit consider which skills you have used and improved, and add them to your reflective log or CV.

21.6 References

Lee, P. and Stuart, M. 1997 Making a Video, *Journal of Geography in Higher Education*, Directions, **21**, 1, 127–134.

Lewis, S. and Mills, C. 1995 Field Notebooks: a student's guide, *Journal of Geography in Higher Education*, Directions, **19**, 1, 111–114.

and see references on safety on p.181.

Lost the plot 2

Use two straight lines to locate four icons per plot. Answers on p.269.

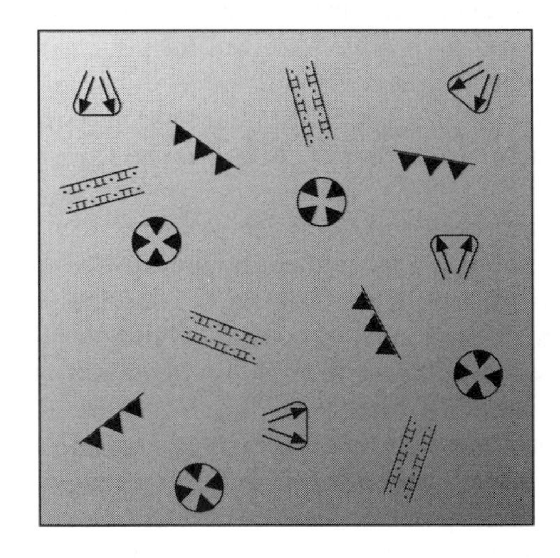

22 PRESENTING GEOGRAPHICAL INFORMATION

A graphics program or sketch will move mountains.

Maps and diagrams usually save lots of words, they are good value especially in exam answers where time is short, and they give implicit evidence of geographical thinking. Plotting, sketching or mapping geographical information exemplifies and clarifies relationships.

Throughout your geographical career people have hounded you about labelling, scales, legends, keys and titles because every map and diagram must have them. So why stop adding them at university? No one will bother to chase you for them, or indeed mention it. Staff will, however, deduct marks for their non-appearance. This is not because life is grossly unfair, which it is, but because all geographers label maps and diagrams correctly, don't they! The skill here is to give careful consistent attention to detail and completeness. Don't forget to acknowledge sources with references as in *'Figure n, This is the title of my diagram (after Penman and Wipe 2010)'*.

There are deliberate mistakes on most of the diagrams in this chapter, so exercise your critical skills. Look at each to see where you would do better.

22.1 FIELD OR PANORAMA SKETCHES

Tragically you may find yourself walking happily in Majorca or Morecambe, when some sad academic says brightly 'Please do a field sketch of the salient urban / cultural / geomorphological / biogeographical / ... features'. Then sits down and starts eating sarnies and everyone mutters about the unreasonableness of the task and not being Fine Art students. These are statements every geography lecturer has heard before, and will be deeply unimpressed to hear again. Sketchers are easily discouraged, but even a rough sketch can convey considerable geographical evidence, which makes them valuable elements in a report or essay. Figure 22.1 is not particularly good, being part sketch and part cross-sectional plan. There is considerable room for artistic improvement, but it will get good marks because it picks out the *evidence* for slope movement through specific examples. It also shows a 'wider view' than a camera. Five photos would be required to show all these points. The sketch includes evidence which the camera could not 'see' on the other side of cars and trees.

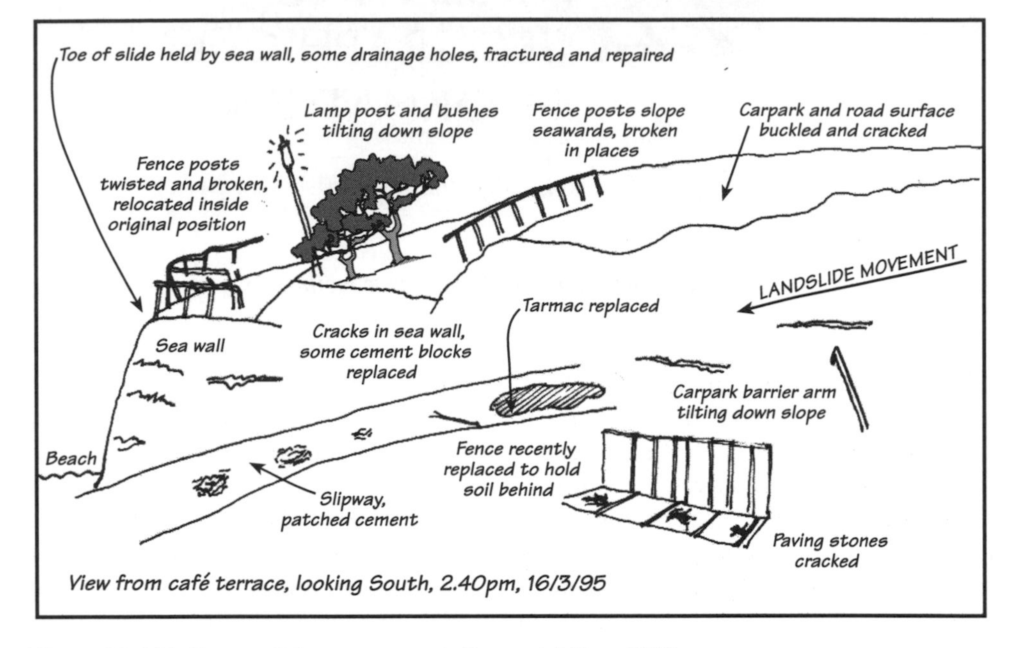

Figure 22:1 Evidence of slope movement, Runswick Bay, 1995

No one ever did field sketches straight off, they take practice, but not much. The trick is to pick out the two or three lines that will anchor the sketch, usually the horizon, maybe a building or two, and then place the geographical features you want to illustrate.

- ✔ Pick out the features 'that matter for this exercise only'.
- ✔ Do not attempt to draw the landscape, it is a waste of time (unless you are Constable reincarnated). Keep it simple.
- ✔ Tools include a firm board, paper, 2B and 4B pencils, pen, rubber and perhaps, binoculars.
- ✔ Locate a couple of 'landmarks' to anchor the sketch, but then simplify.
- ✔ Show general outlines for buildings or slopes but omit details.
- ✔ Shading gives a feeling of depth and distance, turn the pencil sideways. Use thicker, darker lines for close objects and finer lighter lines for those in the distance.

- ✔ Everyone exaggerates the height of objects like hills and buildings to start with. Redraw reducing heights by half to two thirds and compare the result.
- ✔ Shadows clutter a sketch, they are a transient element, a function of the weather and time of the visit, and best omitted.

A field sketch needs to make geographical points. An artist, sketching in the field, covers his or her work with notes on colour, shading and lighting effects. A field sketch needs to be covered with notes that score points in a report. If really suffering from lack of practice, start by doing a field sketch for meteorology! Skies being generally empty and clouds having simple shapes, this is easier (see Fig. 22.2). The clouds are uncluttered by background, and the only landscape element required is an horizon line. Go for it!

Minimum labelling would involve naming the features, in this case the types of clouds, BUT there is an opportunity to give a fuller explanation. You could add information about cloud elevation and whether there will be rain soon. Look at the labels on Figure 22.2, are they correct? What other information might be added? Be critical.

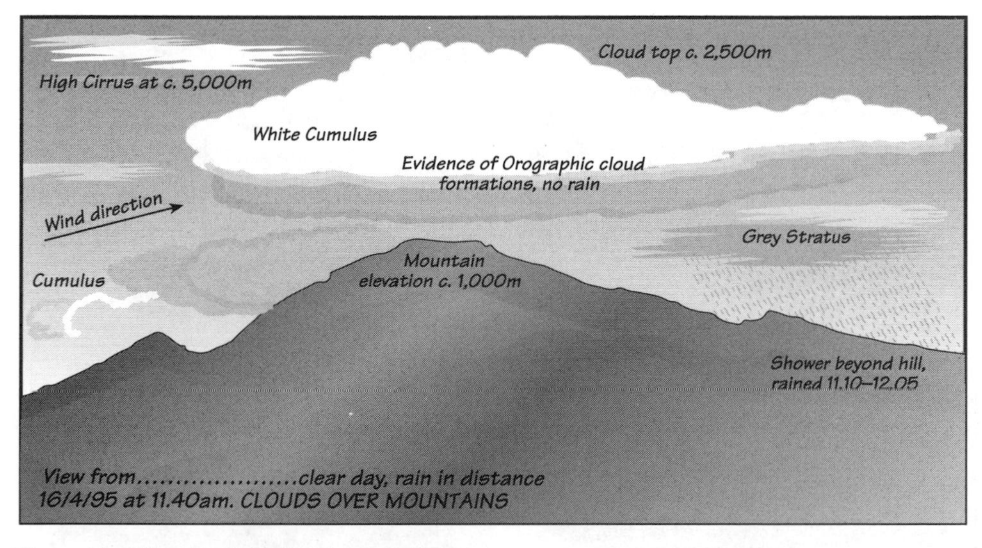

Figure 22:2 Sketch of clouds, April 1995

With townscape and landscape sketches pick out the geographical features you want and ignore the rest. Figure 22.3 shows four sketches of the same site, drawn to illustrate four different geographical aspects of the landscape. You might want to inset a small map to locate the view, or cite the grid reference of the sketching point and compass bearing. Finally, make sure your notes on the map or sketch are legible. But are these sketches? They were sketches, which were scanned into a graphics program and the shading was added. This may look more 'professional', but the hand-drawn version is perfect for field reports and dissertations.

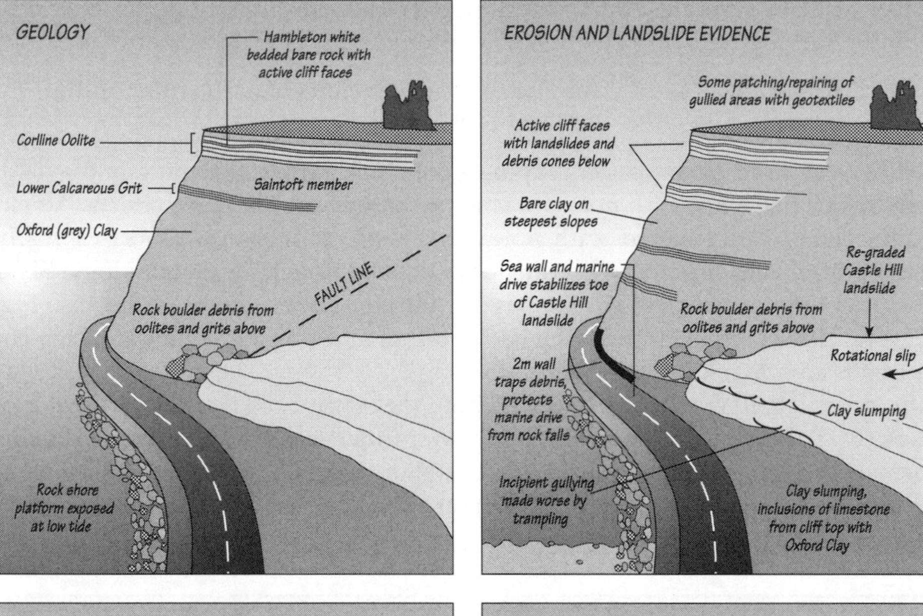

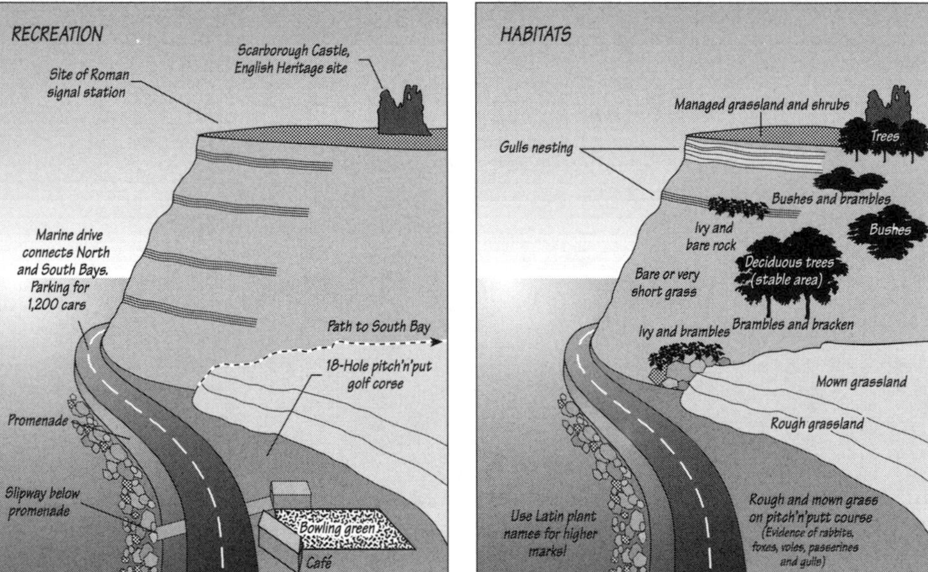

Figure 22:3 Geographical sketches of Castle Hill, Scarborough, North Yorkshire

Cheating – so take photographs and trace it if totally desperate BUT you still need to make lots of notes during your field visit to remind you which features to trace. It is cheaper, quicker and infinitely less hassle to do this task in the field and get rid of it, than to waste time and drinking money on film, developing costs and being delayed by developing time.

22.2 Field plans

Figure 22.4 shows plan sketches, with graphical enhancements, of sections of an urban stream. They make multiple points about flow, vegetation and debris. An indication of SCALE is vital, this is a small tributary not the Mississippi, one diagram has a scale bar, on the other two the width of the stream is shown. In the final version, replace the plant names with their Latin names (see p.240). Good examples of field plans and sketches can be found in RSPB *et al.* (1994).

22.3 Sketch maps

Can you tell me how to get to Sesame Street?

Sketch maps are tremendously valuable in highlighting specific geographical evidence. Either sketch freehand or use a published map as a base. Annotate the map to make the geographical points you require. Again remember to make very good field notes to avoid a second site visit. Unless there is a good reason, put north at the top of the map and always add a direction arrow and indication of scale. Sketches can be scanned into a graphics package and legends added digitally, but hand annotation is fine.

Simplification is the key. A stylised, topologically incorrect map, like the London Underground Map where the spatial relationships between stations are not geographically correct, can clarify points. Figure 22.5 shows a simplified, stylised sketch map of the riverside area of Leeds, picking out some examples of urban regeneration. It is partial (all roads are omitted included a big interchange and the M1), inaccurate cartographically, but conveys evidence for recent urban change. Another researcher would pick different examples of regeneration, the author is not aiming to be comprehensive, but to convey specific geographical information. The sketch is 'anchored' by four landmarks, the station, Tetley's Brewery, the river and railway. There is no precise scale bar because the map is topologically inaccurate; how could you indicate scale more generally?

It is logical to use or adapt conventional signs from Ordnance Survey maps in sketch mapping. If it is a black and white sketch, some adaptation will be necessary, for example to distinguish a road from a river. Symbols used in geomorphological mapping (Cooke and Doornkamp 1974; Gardiner and Dackombe 1983) or architectural drawings (find local documents with a keyword search using architecture, drawing, design, graphic) may also be useful. Add elevation details from survey maps if relevant.

You may be asked to draw a sketch plan or map as an examination question. Here is a sample from recent Leeds University papers: 'Provide an annotated plan of a new retail park, with brief justification for the features you include', 'Sketch and annotate diagrams to explain the kinds of equilibrium that may exist in

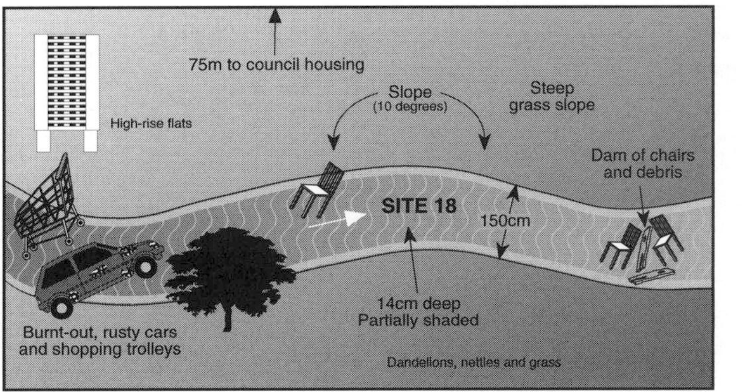

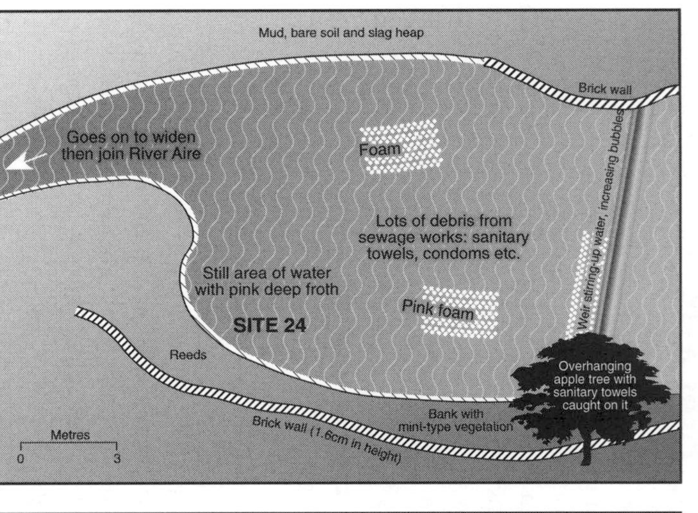

Figure 22:4 Field plans of stream reaches

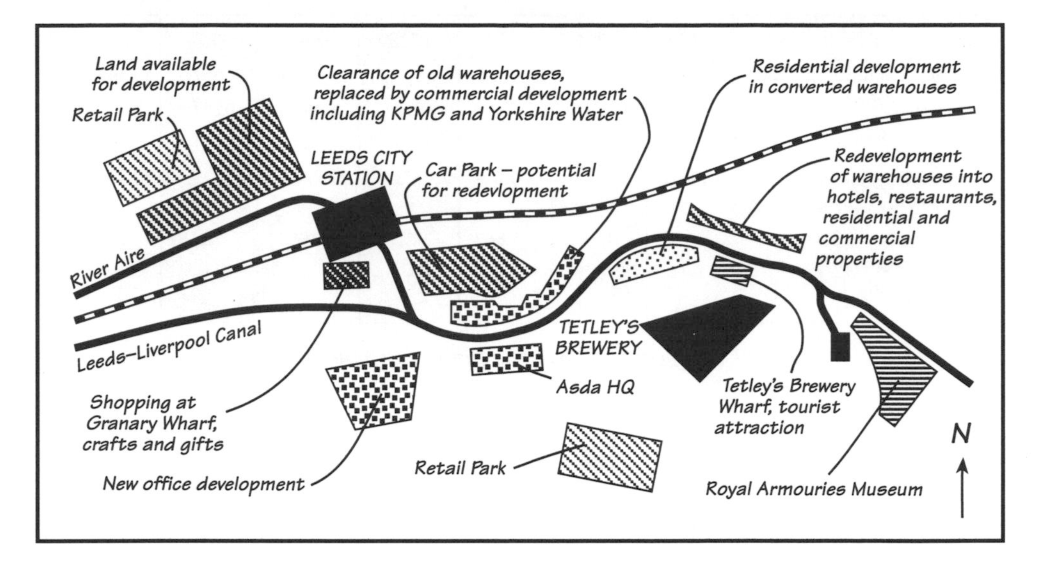

Figure 22:5 Sketch map showing selected examples of urban regeneration along the River Aire, Leeds, 1998

simple models', 'Sketch an outline flow diagram for an ecological simulation model', 'Sketch the micro-depositional forms you may find in alluvial channels'. These were all set as short answer questions, but of course you would practise such diagrams for inclusion in essays too.

22.4 Tables

Most word processing packages can import tables and graphs from database and statistical packages to brighten up your report or essay. This is not a substitute for geographical analysis, a smart graph is not a substitute for astute analysis – but then it never did any harm either.

Things to avoid! avoid! avoid! Look critically at Table 22.1.

Table 22.1 has correct data, but it is awkward to read across the lines with this background, and it is not properly referenced. It might be important to you to highlight the difference between the tourist numbers and US$ data. Is the style in Table 22.2 better? Note the inclusion of the information source in the figure caption. This makes it clear to the reader that this is not your own, primary data but from another source. Does 'Item' in the caption line help – did you notice it?

Item	1980	1985	1987	1988	1989	1990	1991
Total no. of tourists (1000)	5,702.5	17,833.1	26,902.3	31,694.8	24,501.4	27,461.8	33,349.8
Foreigners	529.1	1,370.5	1,727.8	1,842.2	1,461.0	1,747.3	2,710.1
Overseas Chinese	34.4	84.8	87.1	79.3	68.5	91.1	133.4
Hong Kong, Macao and Taiwan Chinese	5,139.0	16,377.8	25,087.4	29,773.3	22,971.9	25,623.4	30,506.3
Total foreign exchange income from tourism (US $10,000)	617	1,250	1,862	2,247	1,860	2,218	2,845

Table 22.1 China: tourist numbers and foreign exchange income from tourism

Item	1980	1985	1987	1988	1989	1990	1991
Total no. of tourists (1000)	5,702.5	17,833.1	26,902.3	31,694.8	24,501.4	27,461.8	33,349.8
Foreigners (1000)	529.1	1,370.5	1,727.8	1,842.2	1,461.0	1,747.3	2,710.1
Overseas Chinese (1000)	34.4	84.8	87.1	79.3	68.5	91.1	133.4
Hong Kong, Macao and Taiwan Chinese (1000)	5,139.0	16,377.8	25,087.4	29,773.3	22,971.9	25,623.4	30,506.3
Total foreign exchange income from tourism (US $10,000)	617	1,250	1,862	2,247	1,860	2,218	2,845

Table 22.2 China: tourist numbers and foreign exchange income from tourism (CSICSC, 1992)

22.5 GRAPHS

Points about clarity of style and presentation are as true for graphs as for tables. In addition, watch out that axes labels are full and explanatory. Figure 22.6 is an example of a graph with scales on both axes and a key but there is no title, the y-axis 0–90 scale is not labelled and has no units, and so the graph fails as an example of good practice.

Is Figure 22.7 any more successful? There is a figure title, the y-axis scale is 0–100 but is still not labelled, are these numbers of birds or percentages, and were they observed in a week, month or year? Marginally better, but still a FAIL.

It is always a sound idea to indicate the error or bias in data, it is often crucial information when drawing inferences. Add error bars to graphs where the main plot is of mean, median or modal values. They indicate the range and therefore the precision of the data. Figure 22.8 illustrates the difference between a graph with and without the error bars. Graph B is much more useful, it shows there is considerable variability hidden by the average. Is the y-axis appropriate? Why 0–150, should it stop at 100?

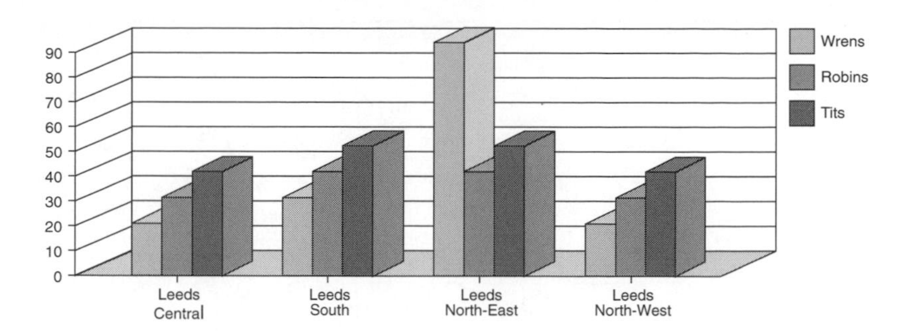

Figure 22:6

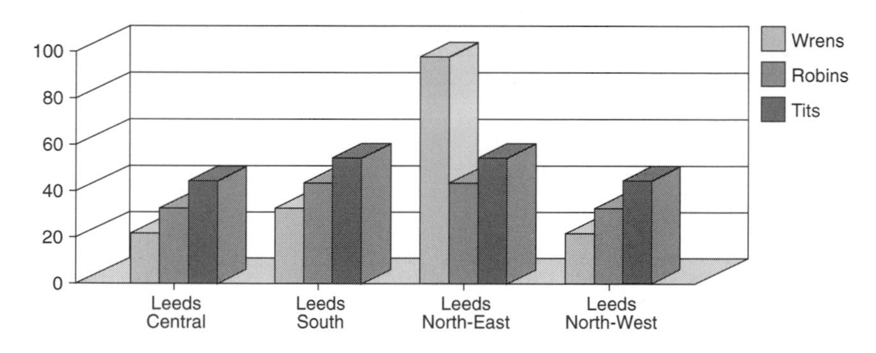

Figure 22:7 Distribution of selected garden birds in Leeds

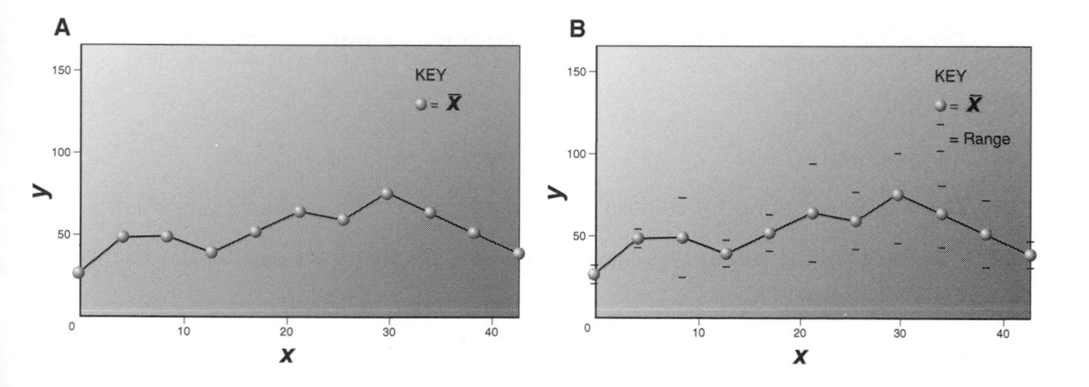

Figure 22:8 Are error bars helpful?

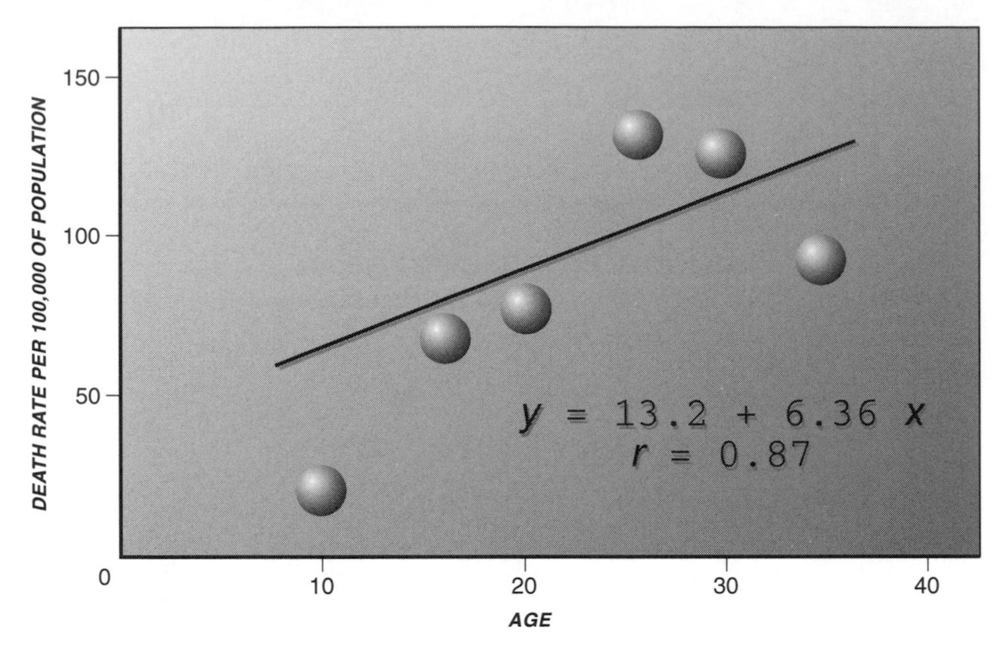

Figure 22:9 Death rates by age cohorts

Figure 22.9 shows a scatter plot with the regression equation for the line and r value added, but look at it carefully. Is this graph OK? It looks smart, and including the regression line and r value is the right thing to do, BUT:

✗ When x = 0, does y = 13.2? Extend the regression line to the y-axis and check. Oh dear, where did this equation come from?

✗ Does the line fit the data? Is there a balanced distribution of points above and below the line? In fact the regression equation is right, it is the regression line that is drawn incorrectly. Oh dear.

✗ There is no citation in the title acknowledging the source of the data.

Good points:

✔ It looks smart.

✔ There is an equation for the line and correlation coefficient, and it is correct.

✔ The line is in the same range as the data, it is not arbitrarily extended to the margins of the graph.

22.6 To table or to graph? That is the question (as someone once said)

Deciding whether to present data on a table or graph may mean experimenting to see which approach gets the message across clearly and without clutter. The following graphs are examples of inappropriate tables and graphs. Endeavour to avoid the errors they exemplify. What is the problem with Table 22.3?

Morbidity rates (all causes) associated with dinghy sailing water exposure in 1990	*Cheddar Reservoir*	*Chew Valley Lake*	*Spinnaker Lake*	*Plymouth Sound*
Number reporting health problems	19	31	23	32
Number reporting no health problems	39	88	51	97
Total	58	119	74	129
Percent	33	26	31	25

Table 22:3 Health risk assessment of dinghy sailing in Avon, UK (Newman and Foster, 1993)

The data in Table 22.3 are taken from the *European Environmental Handbook* (Newman and Foster 1993) although the original source is a paper in the *Journal of the Institute of Environmental Management*. Generally you should reference the source as you viewed it (Newman and Foster 1993). BUT is this one you need to check before quoting further? Cheddar and Chew Valley are likely to be in Avon, but Plymouth Sound has Devon and Cornwall overtones. The Newman and Foster (1993) table title is reproduced here exactly, as in their text, but is it right? If you quote this as Newman and Foster might you be perpetuating an error? Additionally, do you like the table background and should the total and % lines be highlighted differently for greater clarity?

Figure 22.10 is a classic example of how not to present a graph! Mixed variables, numbers and % on the y-axis, no y-axis legend, lack of quality in the x-axis titles, and there is no figure caption. This chart FAILS.

Pie charts can be very useful, but Figure 22.11 is unhelpful where the percentage values for each segment are similar or where there are many fine slices. There is no figure caption. ANOTHER FAIL.

It is hard to imagine something worse than Figure 22.12, although rescaling has made the diagram more legible. Joining the dots implies there is a causal connection between the x-axis variables, the y-axis label has no units, there is no figure caption and, most disastrously, the data mixes whole numbers and %. FAIL WITH BLACK STARS. With the data in Figure 22.13, on the other hand, you can use the 'join the dots' technique, because there is a logical year to year link. Both axis are labelled, this graph is a PASS.

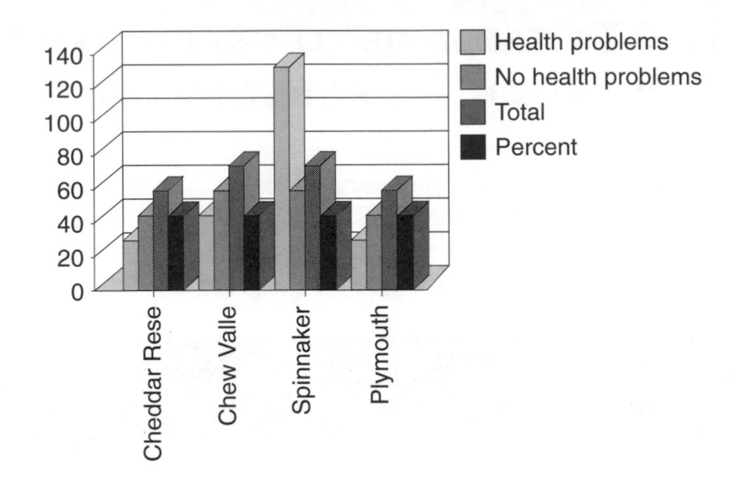

Figure 22:10

Figure 22:11

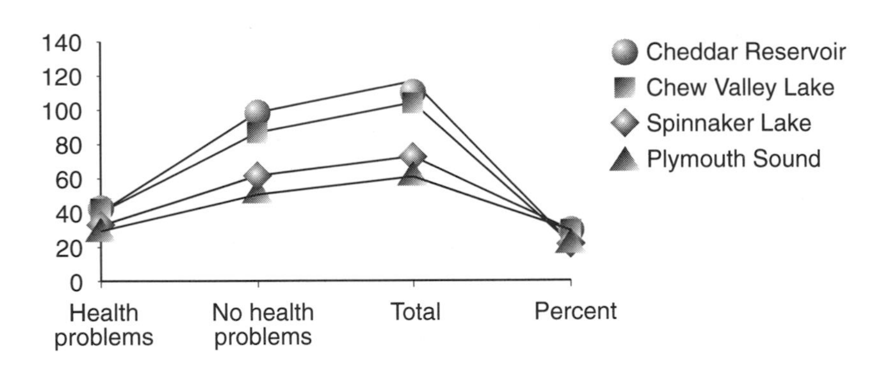

Figure 22:12

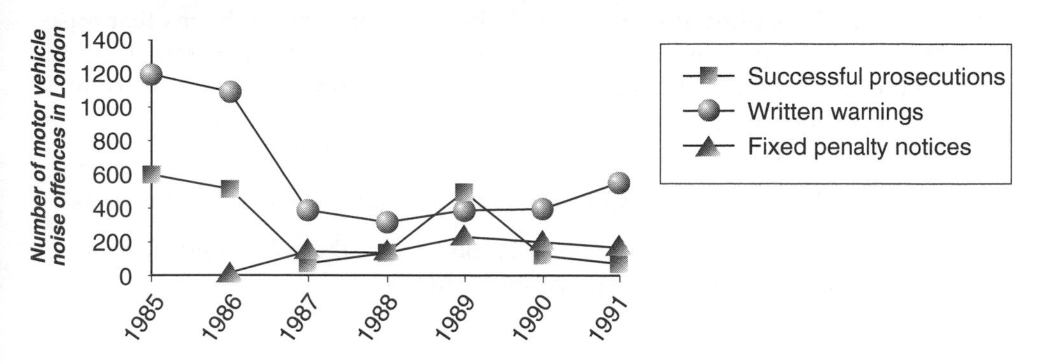

Figure 22:13 Motor vehicle noise offences in London, 1985–1991 (DOE, 1995)

Now pretend you are an examiner and interpret the graphs in Figure 22.14.

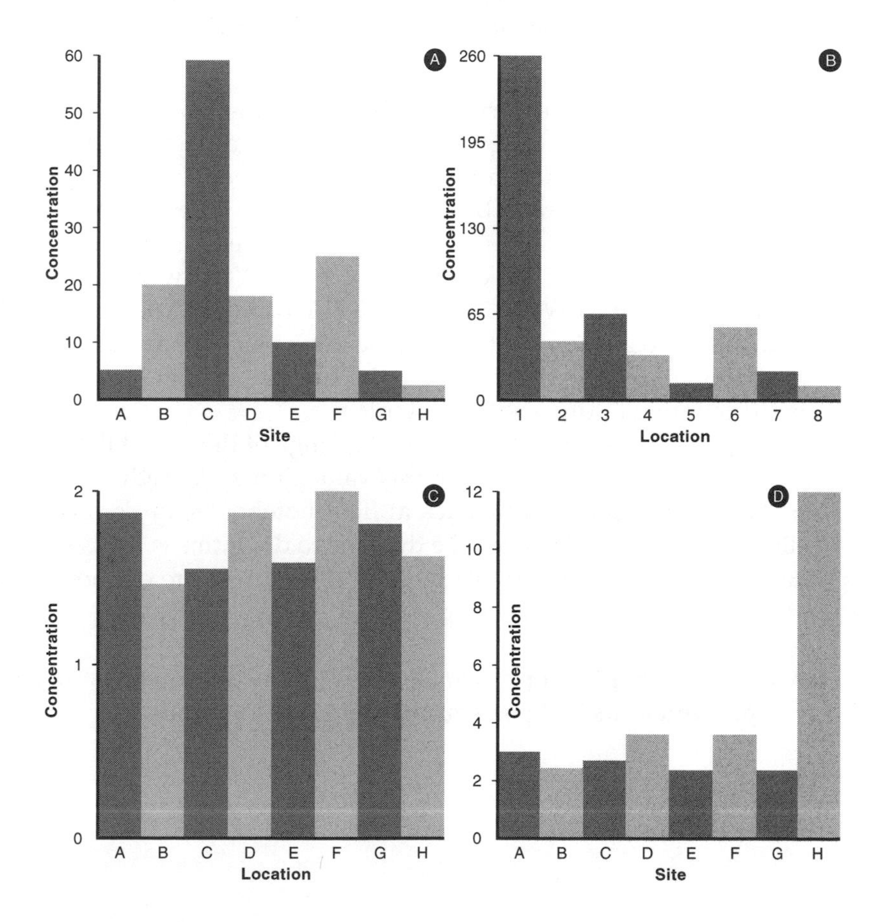

Figure 22:14 Comparing graphs

Have you really looked at Figure 22.14?, then read on. The problems that arise with this figure appear in many dissertations and reports. The results from measurements at H sites on 4 separate occasions, or repeat experiments, or at 4 sites on H occasions are presented. Printing on 4 separate pages often compounds comparison and interpretation problems. Problems are as follows:

✗ the y-axis label is there, but 'Concentration' of what and where are the units?

✗ the x-axis changes from A-H to 1-8 in graph B; were these different sites and nothing to do with the sites in graphs A, C and D?

✗ the x-axis label changes from Site to Location and back again.

✗ using A–D for the graph labels and A–H for the sites is likely to lead to confusion: it is not wrong, but can add confusion.

HOWEVER, these are minor quibbles compared to the y-axis values:

- Graph C appears to have lots of high values, but are the lowest of all. Graph B looks as if most of the values are low, apart from site 1, but they are the highest observations.

This presentation problem arises primarily because some computer packages scale the y-axis according to the range of the data. This is a potential disaster for 'compare and contrast' purposes. Graph C, where the highest value is less than 2 is scaled to fill the page as is Graph B where the highest value is 260. Avoid the problem by scaling the y-axis to a common height, or by exporting the data to a package which allows scaling of the y-axis to the values YOU want, or by hand-drawing the graphs (WHICH WILL BE PERFECTLY ACCEPTABLE). Where the reader needs to compare graphs, put them on the same page. Photo-reduction and pasting a number of graphs on a page is often the efficient option, saving hours of unproductive, frustrating fiddling with printouts.

A value like 260 in Graph B is so far outside the range of the rest of the data that it must prompt a check on the accuracy of this value, it may be right or should it be 26 or 2.6? This may mean looking back at field notebooks, or databases and takes time, BUT it is a job you have to take the time to do; if this is a mis-entry the geographical interpretation changes totally. Data entry errors are ominously easy to make and require very careful checking.

The advice in this chapter applies to all essays, reports, projects, slides for presentations and posters, not just to dissertations.

22.7 References and Further Reading

Cooke, R.U. and Doornkamp, J.C. 1974 *Geomorphology in Environmental Management*, Clarendon Press, Oxford.

CSICSC 1992 *China Statistical Yearbook 1992*, Fan Z., Fang J., Liu H., Wang Y. and Zhang J. (eds) China Statistical Information and Consultancy Service Centre, Beijing, T15.15.

DoE 1995 Motor Vehicle Noise Offences in England, *Digest of Environmental Statistics*, Department of the Environment, HMSO, London, No 17, D26, p.186.

Gardiner, V. and Dackombe, R.V. 1983 *Geomorphological Field Manual*, George Allen and Unwin, London.

Newman, O. and Foster, A. 1993 *European Environmental Statistics Handbook*, Gale Research International Ltd, Andover, UK, Table 278, 151.

RSPB, NRA and RSNC 1994 *The New Rivers and Wildlife Handbook*, The Royal Society for the Protection of Birds, Sandy, Bedfordshire.

Geojumble

Three items in the upper diagram do not appear in the lower one, where there are four new items. Can you sort the jumble? Answers on p.269.

23 POSTERS AND STANDS

Is it possible to organise a riot?

Posters are one of the ways researchers share results at seminars and conferences. They may be part of a passive presentation, or more inter-active, like stands, when the authors are available to answer questions. With both stand and poster presentations the key skill is communication, getting a message across clearly and concisely.

23.1 Posters

Posters have limited space, so the presenter is forced to concentrate on the essential elements, and express them creatively through brief, concise statements and explanations. There is no room for flannel. Most geography departments display the posters produced by staff and postgraduates for conference presentations; corridor, office and laboratory walls are likely sites. Take a critical look at them. How effectively is the 'message' communicated? Are the main points readable at 1–2 m? Are you enticed into going closer and reading the detail? Do you like the colour combinations? Is there too much or too little material? Is there a good mix of pictorial and textural information?

First, check the presentation guidelines. There are often limits on size (which relate to the size of display boards and giving everyone a fair share of the space). Then find out about the audience. Whether it is children, fellow students, an oldies outing or a company presentation, each would benefit from a tailored design and presentation, even through the basic information would be the same.

One of the fastest ways to lose marks is to overload with information. Printing an essay in 14pt font and sticking it on card will lead to instant failure, no matter how good the academic content. Sound bite length messages are wanted. However, this is still an academic exercise, so a sound-bite alone will not do. The academic argument and evidence is required on a poster, as in an essay. Figure 23.1 expresses visually what might in an essay be expressed as: 'In the period 1981 to 1996, motoring costs have doubled. This rate is slightly less than the general rate of inflation. In the same period the cost of coach and rail travel has exceeded inflation rate, increasing 2.5 times in the same period (Church, 1997)'. In Figure 23.1 the evidence is in the graph.

Some attention should be paid to getting a 'grabbing' headline to encourage people to read further. The general advice is to go for simplicity and impact. Consider using a question and answer format to draw the reader into the topic.

Jolly shaped posters, a supertanker for international trade, a greenhouse for

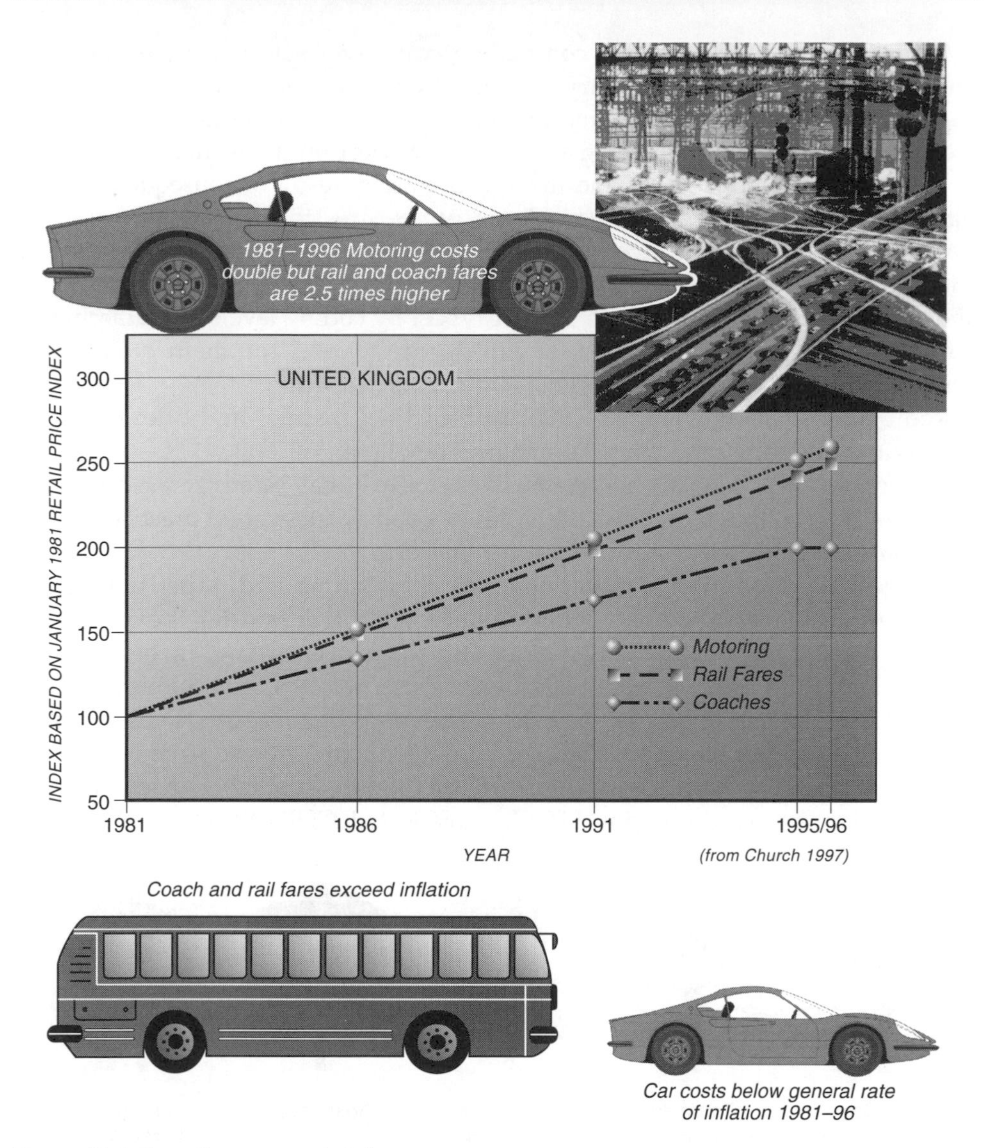

Figure 23:1 Keep the message brief

global warming or a county shape for local planning policy will attract attention. Be careful not to let the background overwhelm the message, the background should complement and enhance, not dominate. It is also important when choosing a background, not to bias the message with an inappropriate, stereotypical image (Vujakovic, 1995). Amongst a gathering of 50 posters on 'Ecology and Sustainability' there are likely to be 48 on green boards with green, cream and brown mounts. Using a different background might make your poster stand out. Whatever the shape, ensure the maximum width and height is within the maximum size guidelines.

Mount pictures and text onto contrasting coloured card or paper to highlight them. You could put primary information on one colour paper and supplementary information on another. Being consistent in design format assists the reader, for example by placing argument or background information to the left of an image, graph or picture, and the interpretation, result or consequence to the right. Using two different paper colours or textures to distinguish argument statements from consequence statements will reinforce the message. It may be effective to have a hierarchy of information with the main story in the largest type, and more detailed information in smaller types. By coding levels or hierarchies of information consistently, the reader can decide to read the main points for a general overview, or to read the whole in detail as desired.

Remember to acknowledge sources and add keys, scales and titles on maps, graphs and diagrams, and put your name somewhere.

Posters can do a 'compare and contrast' exercise. It may be advantageous to use creative visual coding to indicate examples of good and less good practice (*see* Fig. 23.2), but remember to include a key somewhere.

There are costs involved in poster production, including card, paper, photocopying, enlarging and printing photographic material. Colour printing is expensive so be certain everyone is happy with the size and shape of each diagram before printing. A rough draft or mock-up, in black and white, before a final colour print will save money. The first five to seven drafts are never right; font sizes that seem enormous on a computer screen look small on a poster board viewed from two meters away. Any poster produced by a group will be the subject of much discussion and change before everyone is happy! SPELL CHECK ALL COPY BEFORE PRINTING.

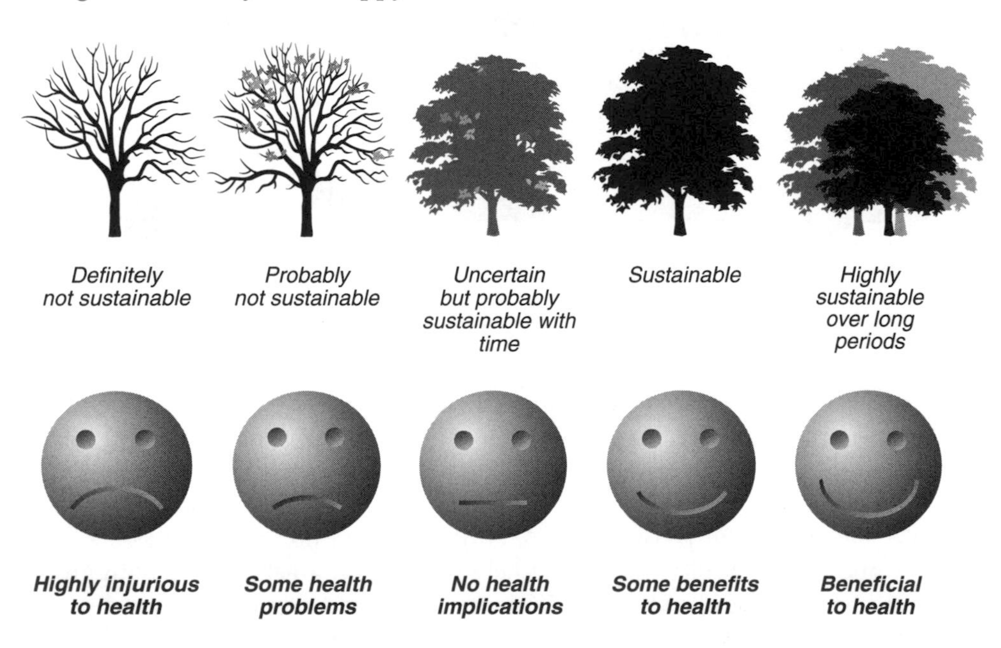

Figure 23:2 Examples of visual codes or classifications

Interactive posters with pop-up book effects, wheels to spin, fish to move as they migrate along rivers, or nests of overlays that slide away can be very effective. Take care that the structure is solid enough to cope with handling. Use a good glue, reinforce cut edges and ensure the poster is very safely attached to the wall or board. Multi-layered card is heavy and likely to fall off.

Show your mock-up, pilot poster to a few people. Ask them what message it conveys. Is it what you intended? Are they following through the material in the order you wish? Can they distinguish major points from support material? What do they feel about the colour scheme? Are the font sizes large enough? Does the overall effect encourage them to read further? Are the main arguments supported by clear evidence? Decide whether the design promotes the delivery of the message and then do the final version. If the poster is to be assessed, self-assess using your departmental assessment form or Figure 23.3. How are the marks are assigned? If the marks matter, change the design to meet the criteria for a higher grade.

23.2 Virtual posters

Virtual Posters are designed electronically in word processing or graphics packages. They have the advantages of being low cost, easy to store and can be viewed via the www. A virtual poster may be an electronic version of a physical poster and the guidelines above apply in the same way. Think about size, the maximum size will probably be defined as part of the project, probably 1–3 screens. Use an attractive and clear typeface and font size. It may be appropriate to match font types to the material, for example antique scripts for historical geographical items or pollen diagrams. Think about whether you find CAPITALS EASIER TO READ or a Mixed Case style. Use different typefaces, colours and shading to highlight different types of information.

A virtual poster can be more fun IF it is possible to include electronic elements. Add HTML links to web sites elsewhere, video, photographs, maps, animated cartoons, short interactive computer programmes to demonstrate a model in action, ... There are many possibilities. The facilities and the expense of scanning material into digital form will govern your creativeness. Beware limits on file size, which may be set as part of the exercise.

A www search in August 1998 found a range of posters from academic courses and conferences, different search engines located different sites. Search using posters+virtual or virtual+posters+university to view current examples.

Group Names .... Poster Title						
	I	2:1	2:2	3	Fail	
Poster Structure (20%)						
Well Organised						Disorganised
Single Topic Focus						No clear focus
Research and Argument (60%)						
Main points included						No grasp of the main issues
Points supported by evidence						No corroborating evidence
Good use of relevant examples						Lack of examples
Ideas clearly expressed						Muddled presentation
Design and Presentation (20%)						
Creative layout and design						Poor design
Good Graphics						No/poor graphics
Good Word Processing						No/poor word processing
Key points readable at 2 m						Key points unreadable at 2 m
Comments						

Figure 23:3 Poster assessment criteria

23.3 Organising an exhibition stand

You may be asked to produce an exhibition stand as part of a module assessment, for a department or university open day, or for a University Society as part of Freshers' Week. Given the effort and time required to produce a good stand it should be a team activity, and will benefit from the fusion of ideas. With five people on a team there will be five different, and all very reasonable, ideas and approaches. Leave time for discussion and consensus, so everyone gets involved and contributes fully. An exhibition involves verbal discussion of research findings with visitors. Conversations of this type require you to adapt the promotional 'blurb' to match visitors' interests and attitudes. This skill is different from an oral presentation for a tutorial or seminar, as it develops the ability to think and adapt the message as you speak.

The nature and experience of the audience need serious consideration. For a module exhibition the audience will include tutors, fellow students and .... WHO ELSE?.... Find out who is coming. Can you assume any background subject knowledge or not? It may be important to decide whether the exhibit will work at a variety of levels to suit all visitors. But remember if you devise a duck hooking competition to attract the kids, the adults will be there too, and keen to hook those ducks. Anything that involves active participation, a quiz, a game, a short video or computer presentation will capture the interest and attention of a visitor.

Find out about the facilities and space available. What is the area and shape for posters? Are there power points for a video, computer, spotlights or other demonstrations? Will there be table and chairs available, so visitors can be encouraged to sit down and discuss issues in comfort?

If you have the chance, take a critical look at an exhibition. Careers Service Fairs on campus are great opportunities, organisations are really trying to sell themselves so need to give clear messages. Consider what materials are used, how they are presented, and decide what you like and dislike, what works and does not work for you. Consider the size of photographs, font size on posters and literature, coloured and multi-coloured backgrounds and the height of materials. What enticed you to certain stands? What encouraged you to talk with exhibitors? Was it the free tea or toffees? What could be adapted for your project?

Use pictures and demonstrations to entice visitors and draw them into the topic even when all the team members are talking to other visitors. Ensure there is a logical arrangement to the material so a visitor can follow through the exhibits in the order you wish. Put posters and the pictures to attract passers-by at eye-level, but pictures to attract younger children should be lower. If you intend people to sit down consider what they can see from that level.

The guidelines for poster production will assist with producing an exhibition stand. Co-ordinating colour and style themes through from posters to video to pc displays to promotional handouts will impress. If this is part of an assessment, the proportion of marks for the presentation are likely to be small, the quality of background research and the quality of discussion with stand visitors will get the bulk of the marks (see Fig. 23.4). Getting the 'story-line' right takes time. Practising your patter on a couple of people before an 'assessor' turns up is a good wheeze. Listening is important, watch for body and language clues and tailor your message to the visitor's interests and experience. The following reflective quotes are from level 2 geographers, who had just completed a two-hour exhibition, talking to a range of people who were not geographers.

The most important thing the stand exercise helped me to learn is ... *'Whilst talking to people new thoughts about skills and proof for them came to you'. 'It is also a skill to talk to people and to have to think on the spot, you have to adapt to different people and questions, you cannot just spurt out the same material'. 'I now know that I need to think*

Stand Title	
Group Names	
Please grade on a 1–5 scale, where 1 is Useless, 3 is Average, 5 is Brilliant.	
1. How well did the group articulate the principle points and issues? Comments	1 2 3 4 5
2. How broad and effective was the evidence supplied to justify the claims? Comments	1 2 3 4 5
3. How well did individuals use personal illustration and anecdote to support the claims? Comments	1 2 3 4 5
4. How well did the stand materials support and enhance the points being made? Comments	1 2 3 4 5
5. How creative were the ideas? How much impact did they have? Which materials or approaches would you commend? Comments	1 2 3 4 5
Additional comments:	

Figure 23:4 Assessment criteria for an exhibition stand

through what I want to say in advance, so the message is really clear'. 'I learned a lot about interview technique, so that when people fired difficult questions at me I can deflect them and try to answer them without looking visibly flustered – therefore very helpful'. 'It was easier to talk to one person than a whole room, as you can adjust to their reactions and feel they are listening'.

I most enjoyed ... *'Creating the actual stand'. 'Getting lots of ideas from the other team members'. 'Seeing how good other peoples presentations were'. 'Feeling a sense of achievement when someone enjoyed looking at the stand'.*

I least enjoyed ... *'Standing around waiting for people to talk to us, feeling like a spare part' 'Being asked quite difficult questions because it was difficult to think on your feet and there was only limited time for research'. 'Some of the people were really difficult to talk to'. 'Talking to the first person, but it got better as time went on'.*

Organising and running an exhibition stand is great fun and very exhausting. Have some chocolate handy!

Answer the question that is posed, not one of your own

23.4 References and further reading

Church, J. (ed.) 1997 *Social Trends 27*, The Stationery Office, London.
Vujakovic, P. 1995 Making Posters, *Journal of Geography in Higher Education*, Directions, **19**, 2, 251–256.

Geo-cryptic crossword 2

Answers on p.269.

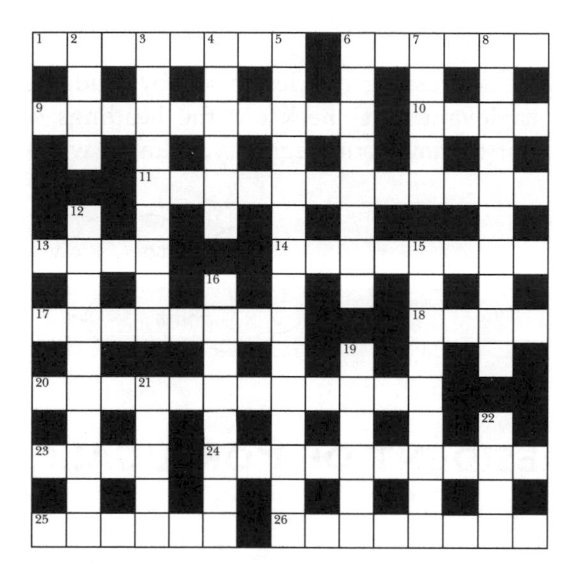

Across

1 Fell under stress, fell (8)
6 Discordant tenor I direct (6)
9 At rock bottom on acid (10)
10 An audible barrier around marshes? (4)
11 It would be a cliché to put into a source of copper pyrites (12)
13 Sounds like you should chase off a cabbage or cream bun (4)
14 He made extravagant gestures; found downwind (3, 5)
17 Excited early view of earth rotation (4, 4)
18 Take in, out, to ignite a French holiday home? (4)
20 Headless pixie, we hear and Alec turn somersaults over terminological definition. (12)
23 Humped ox in prize bush! (4)
24 Could play if suspension is not too long (10)
25 Scanning the lines could lead to arrest (6)
26 In earlier times a vile mend drunk banquets (8)

Down

2 Capable, and incapable (4)
3 A coating when led away from the druid cluster (9)
4 Lain by alien in North Africa (6)
5 After the settlers rule was over? (15)
6 Possessor of love, a commander in chief and acceptable support (8)
7 Conclude an artful arrangement with Fern I did (5)
8 One short of a score in the clubhouse (10)
12 Ply lax hero with wine, makes it winter on the vine! (10)
15 A collection of gravel (9)
16 Long key to the heavens above and inn below (5, 3)
19 Might be carpeted in difficult country (6)
21 Attack Sally (5)
22 Rainbow gala in brown, blue, green or both, but just the one (4)

24 'OF SHOES, AND SHIPS, AND SEALING WAX ...' THINGS STUDENTS ASK

Here is a potpourri of topics and lists addressing queries raised by students. Some items may seem bizarre and irrelevant, but check out the headings so that, should something become relevant during your degree, you know where to look.

24.1 WHO IS THE PRESIDENT OF PORTUGAL?

In regional geography modules like Perspectives of Europe, The Geography of International Relations or World Issues, lecturers tend to leap straight into economic, urban, political, historical and social geography issues assuming that educated students have a background working knowledge of the history and politics of the UK, Europe and the World. If you are unsure who's who and who's where, try answering the following. It can be a useful revision game, as is **Try This 24.1**.

- ✔ Sketch the outline of the countries you are studying, adding the two largest cities and main river system.
- ✔ Name the five largest cities in England? Great Britain? the British Isles? France? the US of A and Brazil?

People to Know, name the following. (There are no answers because many answers listed in the summer of 1998 would be out of date by publication. Be sure you have the current information.)

Who's Who Internationally?	Who are the UK's / Your Country's Leaders?
President of the United States of America	Prime Minister of Great Britain
Vice-President of the United States of America	Foreign Secretary
Is the US President a Democrat or Republican?	Home Secretary
Chancellor of Germany	Chancellor of the Exchequer
President of France	
Prime Minister of France	Leader of the Opposition
President of China	Shadow Chancellor
Prime Minister of Japan	First Minister for Scotland
President of Germany	Archbishop of Canterbury
President of Russia	
Prime Minister of Russia	Governor of the Bank of England
Secretary General of the UN	Which political party is in power now?

If you are studying less familiar regions, some pre-module preparation is a good wheeze. Check out the answers to the following generic questions adapting them for your module.

? What are the countries in the region?

? What are the capital cities of each country?

? Who is the President or monarch and/or Prime Minister of each country?

? Is there a regional economic or political grouping, like ASEAN, CARICOM, NATO, OAS, OAU, or OPEC?

? What has caused any of these countries to appear in the news in the past year?

? Sketch the national boundaries and locate the capitals.

? What kind of government is currently in power in each country?

? In democratic systems, which political party is in power and in opposition?

? What are the main religious groups in each country?

? Which web sites give information for these countries?

Why was the President of Sierra Leone in the news in 1998? Some information dates very quickly, for a geo-politics or world issues based module it is vital to have up to date information.

B. Hunter (ed.) 1997 *The Statesman's Yearbook, a statistical, political and economic account of the states of the world for the year 1997–1998*, (134th edn), MacMillan Reference, London.

TRY THIS 24.1 – The summer of 1998 quiz

Match the people to the office. In 1999 this should be easy, by 2001 some answers will elude you, so update the quiz. Answers on p.269.

1. President of Sierra Leone	a. Daniel Arap Moi	11. President of China	k. Emeka Anyaoku
2. Prime Minister of Eire	b. Nelson Mandela	12. Prime Minister of India	l. Li Peng
3. Secretary-General of the Commonwealth	c. Robin Cook	13. President of South Africa	m. Akihito
4. Emperor of Japan	d. Vaclav Havel	14. Secretary-General of the United Nations	n. Yasser Arafat
5. President of the European Commission	e. Boris Yeltsin	15. President of Brazil	o. Binyamin Netanyahu
6. Prime Minister of Israel	f. Ryutaro Hashimoto	16. British Foreign Secretary	p. Bertie Ahern
7. President of Czech Republic	g. Jiang Zemin	17. President of Russia	q. Jacques Santer
8. Prime Minister of Pakistan	h. Atal Behari Vajpayee	18. Prime Minister of Japan	r. Fernando Cardoso
9. President of Zimbabwe	i. Kofi Annan	19. President of the Palestine Authority	s. Robert Mugabe
10. Prime Minister of China	j. Ahmad Kabbah	20. President of Kenya	t. Nawaz Sharif

24.2 Mature students (NOT TO BE READ BY ANYONE UNDER 20, Thank you)

'They are all so young, and so bright and I don't think I can do this' Oh yes you can. You have had the bottle to get your act together and make massive arrangements for family and work, so handling a class of bright faced nineteen-year-olds is easy. Just keep remembering that all those smart teenagers have developed loads of bad study habits at school, are out partying most nights, are fantasising about the bloke or girl they met last night/want to meet, and haven't got as many incentives for success as you. This degree is taking time from other activities, reducing the family income and pension contributions, which is great motivation for success. Most mature students work harder than students straight from school and do very well in finals.

'I really didn't want to say anything in tutorial, I thought they would laugh when I got it wrong.' I'm nervous, you are nervous, he ... (conjugate to gain confidence or get

to sleep). At the start of a course everyone in the group is nervous. Your experience of talking with people at work, home, office, family, scout group, ... means YOU CAN DO THIS. It is likely that your age gets you unexpected kudos, younger students equate age with experience and are likely to listen to and value your input.

'I hated the first week, all those 19 year olds partying and I was trying to register and pick up the kids.' The first term is stressful, but having made lots of compromises to get to university give it a go, at least until the first set of examinations. If life is really dire you could switch to part-time for the first year, gaining time to get used to study and the complexities of coping with home and friends.

Mature students end up leading group work more often than the average. Your fellow students will be very, very, very, very happy to let you lead each time, it means they can work less hard. Make sure they support you too!

'It takes me ages to learn things, my brain is really slow.' OK, recognise that it does take longer for older brains to absorb new ideas and concepts. The trick is to be organised and be ACTIVE in studying. This whole book should prompt useful ideas. The following suggestions are gleaned from a variety of texts and talking to mature students in Leeds. It is a case of finding the tips and routines that work for you. Like playing the 'cello, keep practising.

Keep fit!

Top Tips

- Review notes the day after a lecture, and at the weekend.
- Check out your notes against the texts or papers; ask 'do I understand this point?'
- Practise writing regularly, anything from a summary paragraph to an essay. Write short paragraphs, which summarise the main points from a lecture or reading, for use in revision. Devising quizzes is also fun (p.188).
- Talk to people about geography, your friends, partner, children, people you meet at bus stops, the dog or hamster. It gives great practice in summarising material, and in trying to explain in an interesting manner you will raise your own interest levels. Non-experts often ask useful questions.
- Try to visit the library on a regular basis, timetable part of a day or evening each week, and stick to the plan.
- Meet a fellow mature student once week or fortnight to chat about experiences and coursework over lunch, coffee or a drink.
- Have a study timetable, and a regular place to study – the shed, attic, bedroom or a corner of the hall. Stick a notice on the door that says something like:

- Be assured that once the family knows you are out of sight for a couple of hours they will find a series of devious things to do unsupervised.
- Remind yourself why you decided to do this degree, then ask is this statistics or computer practical really worse than anything else you have ever had to do? OK, so it is worse, but it will be over in X weeks.
- Some time-management and a bit of organisation will get you through and by the second year you will know so much more about geography and about how to manage than you did at the start.
- When you have finished and got your degree, your family will be amazed, stunned and you will have the qualifications you want.
- Make specific family fun time and stick to it: 'As a family we really tried to go out once a week for an hour, an evening, or all day. This made a real difference. We did walks, swimming, supper at the pub, visits to friends and family and loads of odd things. We took turns to choose where we went, which the kids thought was great.'
- Don't feel guilty, life-long learning is the way the world is going, and there is increasing provision and awareness of mature students needs.

Problems?

Check out the Students Union. Most Union Welfare Offices have advice sheets for mature students, advisors for mature students and a Mature Students Society. If there is no Mature Students Society then start one. Generally, geography staff are very experienced in student problems and in solving them. Most departments have a staff tutor or contact for mature students. Go and bother this nice person sooner than later. If you feel really guilty, buy him or her a drink sometime!

Family management

Even the most helpful family members have other things in their heads besides remembering that they promised to clean out the rabbit, vacuum the bedroom or buy lemons on the way home. On balance a list of chores that everyone agrees to and adheres to on 40 per cent of occasions, is good going. Thank the kids regularly for getting jobs done, and remember this degree should not take over all their lives too. YOU NEED to take time away from study too, plan in trips to cinema, games of badminton, visit to sports centre and get away from academic activities too.

Short courses

If things get really rough it is possible to suspend studying for a semester or year, or stop after one or two years. Most universities offer some certification for completion of each whole or part year. Find out about your options.

Rickards, T. 1992 *How to Win as a Mature Student*, Kogan Page, London.
Wade, S. 1996 *Studying for a Degree: how to succeed as a mature student in higher education*, How To Books, Plymouth.
(Or do a library search using study, degree, mature, as keywords, and check your University Union web site.)

24.3 Dropping out

'I've been here six weeks, no one has spoken to me and I hate it, I'm off.' Happily this is a rare experience for geography students, but every year there will be a few people amongst the thousands taking geography degrees who are not happy. The main reasons seem to be 'wrong subject choice, I should have done ...', 'everyone else is cleverer than me', ' the course was not what I expected', 'I was so shattered after A level and school I really need a break and a rest' and 'I really felt I didn't fit in and it wasn't right'. There are big differences between school and university, homesickness is not unusual. All geography departments lose students for these kinds of reasons, you will not be the first or last geography student feeling unsettled. There are tutors to give advice and people in the Union. Talk to someone as soon as you start to feel unhappy, waiting will probably make you feel worse.

The good news is that most people stick out the first few weeks, get involved with the work and social and sporting activities, and really enjoy themselves. Remember, there are 6000–22000+ other people on your campus, 99.9 per cent are very nice, and at least 99.7 per cent feel as shy as you do. If you are really at odds with university life explore the options of suspending your studies, taking a year out in mid degree or transferring to another university. Take the time to make this a real choice, not a rushed decision.

Question all unsupported statistics!

24.4 Accuracy, Precision, Decimal Places, Uncertainty, Bias and Error!

Most geographical data are noisy, imprecise, inconsistent and may also be biased. The trick is to recognise sources of error, acknowledge them and discuss their impact, for example in the methodology or discussion sections of a report. Discussing error is not a matter of mega *mea culpa*, 'it was his/her fault' statements. Unless exceptional care is taken, items, questions and variables are forgotten, data collection methods influence the measurements and instruments may be inaccurate. The idea is to minimise all possible errors but only the infallible will succeed. Aim to be as objective as possible.

A true value, the absolutely *accurate*, correct value of something can be hard to determine, more often one has the best estimated value and one can describe it in terms of its accuracy and precision. An accurate value will be right or correct measurement. Data are *precise* when there is little uncertainty related to the measurement.

A grid reference position on the ground can be located with increasing accuracy as the scale of the map increases. On a 1:50,000 map a position may be in error by ±32m, whereas from a 1:25,000 map the error may be ±15m. The accuracy of points located from satellite imagery will depend on the height and type of the platform, the spatial resolution of the images, and be biased by topographic distortions and off-nadir angles during scanning. The accuracy of a GPS (global positioning system) will depend on the sophistication of the GPS instrument and the number of satellites within view.

It is good practice to quote a value with its associated resolution, for example with error bars and scatter plots (*see* Figure 22.8 and Figure 22.9), so the reader has a feel for the precision and accuracy of the information. A location may be given as X ± 16m. An average, 72.25%, may be reported with the range 66.3%–78.2%. Temperature data from a thermometer which can, at best, be read to 0.2 of a degree, would have the uncertainty associated with a reading expressed as 8.2±0.2ºC. Measuring temperature with a thermocouple should give greater accuracy, because the measurement can be made to the nearest 0.01ºC. It is therefore a more precise measurement, 8.23±0.01ºC.

However, if the thermocouple consistently under-reads the thermometer, although the data are more precise, they would be less accurate than those from the thermometer. The readings would be precise but inaccurate, due to instrumental error. Similarly if a GPS is not properly calibrated, subsequent readouts will be inaccurate.

Decimal places

A disadvantage of digital technology is that calculations can be reported to many decimal places. You need to decide what is relevant for a specific study or instrument. Avoid the apparent absurdity of 'a family with 2.45678 children', or 'Therefore we conclude the residents of Hobbiton make 3.21776 shopping trips per week to Buckland' or 'Survey results show the population of Stats-on-Sea are 26.3732% Hindu and 56.4567% Christian'.

So how many decimal places do you quote? Think about the accuracy you require and the application. An average temperature of 16.734567ºC is experimentally accurate and appropriate for a physicist doing neutron experiments, for a soil temperature 16ºC or 16.7ºC will do. A digital pH meter used in the laboratory or field may show 2 decimal places, but the digits are unlikely to settle because pH is not a stable measure at this level. In practice, a soil pH is going to have local variability and to report to better than 0.2pH has no practical pedological value. Therefore waiting for an instrument to settle at 6.03 or 6.04 is of no scientific value and a waste of time.

Is the sample biased?

Bias is a consistent error in data. There should be plenty of information in statistics and laboratory and fieldwork sessions about taking samples in the right way and with enough replicates so that bias is minimised. Think about potential sources of bias when considering results. I think you would agree that sampling two mid-day temperatures in July at Filey in 1998, and using these data to describe the average climate of the UK in the twentieth century would leave the reader unimpressed. Such data are obviously inadequate and biased. Does your data set have similar, albeit less blatant, drawbacks?

Data entry errors

These are very, very, very, very common. Check carefully every time, against the original field or laboratory datasets. In statistics packages use the command to show the smallest and largest datum, the commoner errors include forgetting a decimal or typing two items as a single entry.

Computational errors

Have all the formulae for calculations been entered correctly? Aside from cross checking the equations look at the answers and make sure they are in the right 'ball park'. A quick, back of the envelope calculation using whole numbers can give a feel for the range of answers to expect. If answers fall outside the expected range, check all the calculation steps carefully. Keep units the same as far as possible, or translate so that units are the same, and write down the units at each stage of a calculation.

More planning –
more
marks

24.5 Equal opportunities, harassment, gender and racial issues

All universities have equal opportunities policies, they aim to treat students and staff fairly and justly. Happily most of the time there are few problems. This section cannot discuss these issues in detail, but if you feel you have a problem then here are some sources of advice.

Talk to someone sooner rather than later, there will be a departmental tutor who oversees these issues. Your problem is unlikely to be a new one, and there is normally a great deal of advice and experience available. It really is a matter of tapping into it. Check out your Student Union and your University or College handbook. There should be a contact name, someone who has a title like University Equal Opportunities Officer, or Adviser on Equal Opportunities.

Check out your own university web sites, or visit these:

Loughborough University 1997 *Equal Opportunities*, University Handbook Online, [on-line] http://www.lut.ac.uk/admin/central_admin/policy/student_handbook/section13.html Accessed 10 January 1999

Liverpool John Moores University 1997 *Equal Opportunities in LJMU*, University Policy, [on-line] http://www.livjm.ac.uk/equal_op/opportun.htm Accessed 10 January 1999

Adams, A. 1994 *Bullying at Work: how to confront and overcome it*, Virago Press, London.

Banton, M. 1994 *Discrimination*, Open University Press, Milton Keynes.

Carter, P. and Jeffs, A. 1995 *A Very Private Affair: sexual exploitation in Higher Education*, Education Now Books, Ticknall, Derbyshire.

Clarke, L. 1994 *Discrimination*, Institute of Personnel Management, London.

Davidson, M.J. and Cooper, C.L. 1992 *Shattering the Glass Ceiling: the woman manager*, Paul Chapman, London.

24.6 Stress

All students suffer from stress, it is normal, but needs management. Stress is a bodily reaction to the demands of daily life, and arises when you feel that life's physical, emotional or psychological demands are getting too much. Some people view stress as a challenge and it is vital to them for getting jobs done. Stressful events are seen as healthy challenges. Where events are seen as threatening then distress and unhappiness may follow.

If you are feeling upset, or find yourself buttering the kettle and not the toast, try to analyse why. Watch out for circumstances where you are stressed because of:

☹ expectations you have for yourself *(Are they reasonable at this time in these circumstances?)*

☹ expectations of others, especially parents and tutors *(They have the best motives, but are these reasonable expectations at this time?)*

☹ physical environment, noisy flat mates, people who don't wash up, wet weather, hot weather, dark evenings *(What can be done to ameliorate these stresses?)*

☹ academic pressures, too many deadlines, not enough time to read *(Can you use friends to share study problems? Would it help to talk to someone about time management?)*

☹ social pressures, partying all night *(Are friends making unreasonable demands?)*

Serious stress needs proper professional attention, no text will substitute, but how do you recognise stress in yourself or friends? Watch out for signs like feeling tense, irritable, fatigued, depressed, lacking interest in your studies, having a reduced ability to concentrate, apathy and a tendency to get too stuck into stimulants like drink, drugs and nicotine. So that covers most of us, DO NOT GET PARANOID. PLEASE do not wander into the health centre waving this page, and demand attention for what is really a hangover following a work-free term. Thank you.

Managing stress effectively is mostly about balancing demands and desires, getting a mixture of academic and jolly activities, taking time off if you tend to the 'workaholic' approach, (the workaholic student spends nine hours a day in the library outside lectures, has read more papers than the lecturer and is still panicking). Get some exercise – aerobics, line dancing, jogging, swimming, any sport, walk somewhere each day, practise relaxation skills – Tai Chi classes are great fun. Study regularly and for sensible time periods. Break down big tasks into little chunks and tick them off as you do them.

✔ If you are stressed, TALK TO SOMEONE.

24.7 Atlases and dictionaries, do I need them?

Atlases are useful, expensive and the university library should have plenty. An atlas is not a vital purchase, but if you are doing regional geography options you must make the time to check out and learn the names of relevant cities, rivers and regions. For modules on Japan, China and the CIS, this is a non-trivial activity. If your Gran wants to give you something useful for Christmas then an atlas is a good wheeze. Check the edition date, there have been many place name changes in the last twenty years.

It is the consensus of a non-representative random sample of students who have walked into my office in the last week that *you need a general dictionary* and a thesaurus too. Not a geographical dictionary, these are in the library and normally cannot be borrowed so they are always available for reference. Lecturers define technical geographical terms as they go along, so technical vocabulary expands rapidly as you discover new geographical material. It is the non-geographical words that are likely to catch you. You probably don't need a dictionary if you are familiar with:

actuate, adumbrate, antithesis, autonomic, biennial, codex, cointreau, commensurate, consensual, diurnal, emend, enigma, epitome, ergo, esoteric, exponential, extrinsic, facile, fallacious, fecund, flora, genus, gestalt, harbinger, heliocentric, holistic, idiom, induction, intangible, intercalate, interface, interim, isotropic, juxtaposition, lacustrine, latent, leitmotif, lingua franca, locus, logical positivism, macula, mandatory, melange, milieu, minesapint, mutable, palimpsest, paradigm, photoperiod, plutocrat, postulate, quantum, quotient, regolith, retsina, salient, schism, simulation, stigma, stratum and strata, synergy, synthesis, temporal, tequila, transverse, *trompe l'œil*, ubiquitous, Utopia, *vis-à-vis*, viscous, vitreous, xerophyte, zeitgeist

Addenda and corrigenda welcome!

Geographical dictionaries available in a library near you include:

Clark, A.N. 1985 *Longman Dictionary of Geography: human and physical*, Longman, Harlow.

Goodall, B. 1987 *The Penguin Dictionary of Human Geography*, Penguin Books, London.

Goudie, A., Atkinson, B.W., Gregory, K.J., Simmons, I.G., Stoddart, D.R. and Sugden, D. (eds) 1994 *The Encyclopaedic Dictionary of Physical Geography*, (2nd edn), Blackwell Reference, Oxford.

Johnston, R.J., Gregory, D. and Smith, D.M. (eds) 1994 *The Dictionary of Human Geography*, (3rd edn), Blackwell, Oxford.

McDowell, L. and Sharp, J.P. (eds) 1998 *A Feminist Glossary of Human Geography*, Arnold, London.

Mayhew, S. and Penny, A. 1992 *The Concise Oxford Dictionary of Geography*, Oxford University Press, Oxford.

Always use your spell checker but
'spil chqers due knot awl weighs git it write'.

24.8 Latin words and phrases

amos, amas, where is that lass?

Few people do Latin in school now but there are many Latin phrases in normal, everyday, usage. Understanding some of them will get you through university and assist in solving crosswords for life.

a priori	reasoning from cause to effect
ad hoc	for this unusual or exceptional case
ad hominem	to the man, used to describe the case where an argument is directed against the character of the author, rather than addressing the case itself
ad infinitum	to infinity
ad interim	meanwhile
carpe diem	seize the day
CV	curriculum vitae, a short description of your life, suitable for employers
de facto	in fact, or actual
e.g.	exempli gratia, for example
et al.	*et alia*, and other persons, appears in references to refer to multiple authors
etc.	et cetera, and the rest, and never used in a good essay. It implies lazy thinking
et seq.	*et sequens*, that which follows
ex officio	by virtue of office
honoris causa	as an honour
i.e.	id est, that is
inter alia	amongst other things
ipso facto	thereby or by the fact
mea culpa	it was my fault
n.b.	nota bene, note carefully, or take note
op. cit.	opere citato, in the work cited
post hoc	used to describe the type of argument where because x follows y, it is assumed y causes x
reductio ad absurdum	reduced to absurdity
rus in urbe	the country in the town
(sic)	*sic*, which means so, is often printed in brackets to indicate that the preceding word or phrase has been reproduced verbatim, often to point up incorrect spelling
sic transit gloria mundi	nothing to do with Gloria being ill on Monday's bus, it means so passes earthly glory
tempus fugit	time flies
vi et armis	by force of arms
viva voce	by oral testimony, usually shortened to vivas, meaning an oral examination

and finally the motto for many a good fieldclass:

ergo bibamus	therefore let us drink

24.9 Greek alphabet

I thought I understood that rivers stuff, but what is this lamb deer he goes on about?

Physical geographers in particular will find lecturers freely flinging thetas and gammas around in lectures. Duck. The bold items here are the basics. You can write faster using symbols, so it is worth knowing a few of the regular stars.

A	**α**	Alpha	I	ι	Iota	Ρ	ρ	Rho
B	**β**	Beta	K	κ	Kappa	**Σ**	σ	Sigma
Γ	**γ**	Gamma	Λ	**λ**	Lambda	T	**τ**	Tau
Δ	δ	Delta	M	**μ**	Mu	Y	υ	Upsilon
E	ε	Epsilon	N	ν	Nu	Φ	**φ**	Phi
Z	ζ	Zeta	Ξ	ξ	Xi	X	**ψ**	Chi
H	η	Eta	O	o	Omicron	Ψ	**υ**	Psi
Θ	θ	Theta	Π	π	Pi	Ω	ω	Omega

Every water resource scientist knows that ***Eureka*** *means 'the bath water's too hot'.*

24.10 Plant names

Where did houseplants live before urbanisation? No houses, no houseplants.

Biogeography lecturers (and other geographers) have a habit of calling plants by their correct Latin names. This is not because biogeographers are unnaturally pedantic, but because the local or common names for plants change as you move around the country. Be aware that botanists reclassify plants as more is known about their biology and genus, so an up-to-date flora is vital.

All Latin names are indicated by underlining in hand-written text and by *italics* in typescript. When a plant's genus is referred to generally, rather than a particular species, the convention *Carex sp.* is used. If there are lots of species of the same genus then use *Carex spp.* Some plant names are long and it becomes tedious to repeat them every time in full, so the convention is to use the full name initially, *Carex nigra*, and thereafter *C. nigra*. Be careful, this convention refers to the last mentioned plant.

The lists here are for the British Isles, just some of the more common species that may be mentioned. Every biogeographer would produce a different list. For more detail you need a Flora, a reference book with plant descriptions and pictures. You might use these lists in reverse to increase the scholarly tone of an essay by referring to the Latin name of a plant rather than the common name. These short lists are idiosyncratically sorted by vegetation height, low growing herb layer, shrub and then tree level. This is followed by a list for species found in wet places including rushes and ferns, and finally some grasses.

Herb layer, common low growing plants:

Allium ursinum	Ramsons	*Narthecium ossifragum*	Bog asphodel
Bellis perennis	Daisy	*Plantago lanceolata*	Ribwort plantain
Calluna vulgaris	Heather	*Potentilla erecta*	Tormentil
Chamerion angustifolium	Rosebay willowherb		
Circaea lutetiana	Enchanter's nightshade	*Primula vulgaris*	Primrose
Cirsium palustre	Marsh thistle	*Ranunculus repens*	Creeping-buttercup
Digitalis purpurea	Foxglove	*Rumex obtusifolius*	Broad-leaved dock
Drosera rotundifolia	Round-leaved sundew	*Succisa pratensis*	Devil's-bit scabious
Empetrum nigram	Crowberry	*Taraxacum officinale*	Dandelion
	Trifolium pratense	Red clover	
Erica cinerea	Bell heather	*Urtica dioica*	Common nettle
Erica tetralix	Cross-leaved heath	*Vaccinium myrtillus*	Bilberry
Galium saxatile	Heath bedstraw	*Vicia sativa*	Common vetch
Lotus corniculatus	Common bird's-foot trefoil		

Shrub layer plants include:

Cytisus scoparius	Broom	*Salix repens*	Creeping willow
Euonymus europaeus	Spindle	*Sambucus nigra*	Elder
Prunus spinosa	Blackthorn	*Ulex europaeus*	Gorse
Rosa canina	Dog-rose	*Ulex gallii*	Western gorse

Trees, you will get by with common names sometimes, the useful UK species are:

Acer pseudoplatanus	Sycamore	*Picea abies*	Norway spruce
Alnus glutinosa	Alder	*Picea stichensis*	Sitka spruce
Betula pendula	Silver birch	*Pinus contorta*	Lodgepole pine
Crataegus monogyna	Hawthorn	*Pinus sylvestris*	Scots pine
Corylus avellana	Hazel	*Quercus petraea*	Sessile oak
Fagus sylvatica	Beech	*Tilia europaea*	Lime
Fraxinus excelsior	Ash	*Ulmus glabra*	Wych elm
Larix decidua	European larch		

These trees are also frequent stars of pollen diagrams which appear in Quaternary courses, because the larger, tough coated pollen grains of tree species are especially well preserved in peat.

Damp area species include:

Rushes (Juncaceae)	*Juncus acutiflorus*	Sharp-flowered rush
	Juncus effusus	Soft-rush
	Juncus squarrosus	Heath rush
	Luzula sylvatica	Great wood-rush
	Trichophorum cespitosum	Deergrass
Mosses	*Polytrichum commune*	Common polytricum
	Rhytidiadelphus squarrosus	The Biogeographer's favourite tongue twister!
	Sphagnum spp.	Main bog forming genus with lots of species
Ferns (Filicopsidia)	*Adiantum capillus-veneris*	Maidenhair fern
	Blechnum spicant	Hard-fern
	Dryopteris filix-mas	Male-fern
	Phyllitis scolopendrium	Hart's-tongue fern
	Polypodium vulgare	Polypody
	Pteridium aquilinum	Bracken
Bulrush (Typhaceae)	*Typha angustifolia*	Lesser bulrush
	Typha latifolia	Bulrush
Sedges (Cyperacea)	*Eriophorum angustifolium*	Common cottongrass
	Carex nigra	Common sedge
Horsetail (Equisetaceae)	*Equisetum palustre*	Marsh horsetail

Paddling near *Typha latifolia* is a VERY BAD IDEA, bulrushes only grow in water deeper than wellies.

Grasses should (as at all the best parties) be introduced to you individually, but if words like *Poa*, *Festuca*, *Agrostis* or *Holcus* whip past your ears then think grasses. Here are a few:

Grasses (Graminea)	*Agrostis capillaris*	Common bent
	Ammophila arenaria	Marram grass
	Briza media	Quaking-grass
	Deschampsia flexuosa	Wavy hair-grass
	Festuca ovina	Sheep's fescue
	Festuca rubra	Red fescue
	Holcus lanatus	Yorkshire-fog
	Holcus mollis	Creeping soft-grass
	Lolium perenne	Perennial Rye-grass
	Molinia caerulea	Purple moor-grass
	Nardus stricta	Mat-grass
	Phragmites australis	Common reed
	Poa pratensis	Smooth meadow-grass
	Sesleria caerulea	Blue moor-grass

Floras

Fitter, R., Fitter, A. and Blamey, M. 1996 *The Wild Flowers of Britain and North West Europe*, Collins Pocket Guide, (5th edn) Harper Collins, London.
Hubbard, C.E. 1984 *Grasses*, (3rd edn) Penguin Book, London.
Mitchell, A. and Moore, D. 1985 *The Complete Guide to Trees of Britain and Northern Europe*, Dragon's World, Limpsfield.
Phillips, R. 1980 *Grasses, Ferns, Mosses and Lichens of Great Britain*, Pan Books, London.
Stace, C. 1997 *New Flora of the British Isles*, (2nd edn) Cambridge University Press, Cambridge.

and remember, Blewitt, Cramp Ball and Jelly Babies are fun guys to be with!

24.11 'We didn't do this at A-level, or geography at all'

Geography is a big and diverse subject, no one gets to do all of it, at senior school or anywhere else. No problem. Geography has a long tradition in drawing its experts from subjects other than geography. Use your wider knowledge to look at 'geography' issues from other angles.

Top Tips

- Don't panic.
- If faced with modules like economics or climate for the first time go to the lectures and get the general idea of the topic and material. Sort out the focus of the module.
- Doing background reading before a module starts can confuse, but it may be worth looking in local libraries for texts, to pick out the background either before a course starts or during the first couple of weeks. A senior school text is likely to have the essentials with less detail than a university text, BUT you must move on to university level texts. School texts are starting, not end points.
- Take your senior school notes to university, your soil notes may help someone who can help you via their economic notes. People with a maths background can be very popular in statistics modules.
- If you are having difficulty with essay writing, and especially if you have not written an essay in 3 years, find someone who did English, history or economics for A-level and pick their brains about writing. There are people who have excellent English language skills, and can explain how to use semi-colons.
- Ask people which reading they find accessible and comprehensible.

24.12 EQUATIONS AND MATHS!

The squaw on the hippopotamus is equal to the sum of the squaws on the other two hides.

Just a few points but if you have A-level maths ignore this section. The sight of an equation tends to send perfectly normal human and physical geographers into a state of total panic. *This is not necessary*. No one will ask you to re-work Fermat's Last Theorem. You knew all about equations once, it is just a case of letting it flood back. The real problem with university is that the education you were given from 5–16, gets to have some pay off. Unfortunately some things, like mathematics and chemistry, went into a bit of memory that is currently dormant, so it is time to reopen your access routes. If in doubt ask someone with high school chemistry or biology or consult Van Der Molen and Holmes (1997).

An international geographer will be talking to you about the latest excitements in 'property and landlords', or 'sustainability in the Antarctic' with delightful slides to lull you into a sense of security and suddenly diverts to equations. DO NOT PANIC (well, only a little). Recall that scientific notation is just a shorthand. You can do the easy ones:

Water added to a catchment from precipitation, could be accounted for by evaporation loss, evapo-transpiration loss, runoff in the river, and changes in the soil and groundwater stores. (28 words) OR

$P = E+ET+Q+\Delta S+\Delta G$ (not 28 words!)

It is essential to define the elements of the equation, but having done it once, time and space is saved later by not having to write it all out again. Where appropriate, WHICH IS 99.9999 per cent OF THE TIME, the units of measurement should also be included:

Therefore, Darcy's Law for water flow through a soil sample:

$Q = KIA$ where
- Q = Discharge of water though the soil
- K = Hydraulic conductivity of the soil (a function of water content)
- I = Hydraulic gradient
- A = Cross-sectional area of the soil sample

will score 60 per cent, but is better expressed as

$Q = KIA$ where

Q	=	Discharge of water though the soil	$cm^3.s^{-1}$
K	=	Hydraulic conductivity of the soil (a function of water content)	$cm.s^{-1}$
I	=	Hydraulic gradient	dimensionless
A	=	Cross-sectional area of the soil sample	cm^2

for 100 per cent.

Life gets to be more fun when data comes in the form of tables and matrices. Thus a table that lists all the students in your class from Sheila to ...n, and the local pubs, The Orienteer and Compass to …n can be referred to using matrix algebra. Yes, algebra, DO NOT PANIC. What seems to cause confusion is making a distinction between the code used to refer to positions in a matrix and the actual data values themselves. Basically it works like map reading with grid references. If the data in the matrix refers to cash spent by individuals in the pub over a term as here:

HOSTELRY		The Lost Orienteer	The Lorenz and Lowry	The Mattock and Spade	The Hazard and Impact	... n
		j_1	j_2	j_3	j_4	... j_n
Sheila	i_1	£25.50	£16.75	£0.00	£3.00	
Wayne	i_2	£45.70	£0.00	£16.80	£6.50	
Spike	i_3	£52.80	£12.80	£36.45	£22.40	
Liz	i_4	£28.67	£24.87	£16.85	£35.78	
Jenny	i_5	£45.23	£12.60	£21.76	£0.00	
..						
n	i_n					

Each member of the class is coded as $i_i \dots {}_n$ and each of the pubs as $j_1 \dots {}_n$. The total number in the class is $\Sigma_{i=1-n}$ and the total number of pubs is $\Sigma_{j=1-n}$. You can refer to a single cell, the shaded cell for Liz in the Lorenz and Lowry is i4j2. The total amount spent by Sheila is the sum of all the columns j = 1 – n for row i_1. So that's the sum of Σ 25.50 + 16.75 + 3.00 = £45.25p, and Liz has spent £24.87p in the Lorenz and Lowry.

When you are faced with an equation, read it carefully and recall the rules (like solve elements in brackets first). If you are feeling out of control, verbalise the instructions as in the multiple choice below. Which of these verbal instructions are right?

a+15b-e
- a) Multiply b by 15, then add a and subtract e.
- b) The sum of 15 times b and a, minus e.
- c) a plus 15, times b, and subtract e.

$6(x^3+y^2)$
- d) Take the square of y and add it to the cube of x and multiply the total by 6.
- e) Multiply the cube value of x by six, and then add the squared value of y.
- f) Cube x, then square y, add the two values together and multiply by 6.

$2x^2–y^2$
- g) Two times the square of x, minus y squared.
- h) Square x and y, subtract and multiply by 2.
- i) Square x and multiply by 2, and then subtract the squared value of y.

Right answers are a, b, d, f, g and i.

Percentage problems?

50% = ½ = 0.5 33.33% = ⅓ = 0.333 25% = ¼ = 0.25

20% = ⅕ = 0.2 5% = 1/20 = 0.05 1% = 1/100 = 0.01

For further (and much more useful) memory jogging see:

Van Der Molen, F. and Holmes, H. 1997 Maths Help, in Northedge, A., Thomas, J., Lane, A. and Peasgood, A. *The Sciences Good Study Guide*, The Open University, Milton Keynes, 301–395.

24.13 Chemical notation

You play the symbols, and I'll bang the drum.

Soils or water chemistry lectures can come as a shock if you have not cuddled up to the Periodic Table for a while. The two lists below contain the main chemical elements that feature in geomorphological, hydrological and soils modules. Use them as an aide-mémoire. They are simply convenient, internationally recognised abbreviations, and nothing more. Add others as you encounter them.

Symbol	Name	Symbol	Name	Symbol	Name
Al	Aluminium	Fe	Iron	P	Phosphorous
C	Carbon	H	Hydrogen	Pb	Lead
Ca	Calcium	Hg	Mercury	Rn	Radon
Cd	Cadmium	K	Potassium	S	Sulphur
Cl	Chlorine	Mg	Magnesium	Se	Selenium
Cr	Chromium	Mn	Manganese	Si	Silicon
Cs	Caesium	N	Nitrogen	U	Uranium
Cu	Copper	Na	Sodium	Zn	Zinc
F	Fluorine	O	Oxygen		

The knack is to read chemical formulae as sentences. The following sentence is straightforwardly factual:

'Building stone erosion is caused by acidified rainfall, in the form of sulphuric acid, reacting with limestone to form the weaker more erodible, calcium sulphate, carbon dioxide and water'. This next line says exactly the same thing in more scientific notation. The trick is to substitute chemical symbols for words:

$$H_2SO_4 \quad + \quad CaCO_3 \quad \rightarrow \quad CaSO_4 \quad + \quad CO_2 \quad + \quad H_2O$$

(Sulphuric acid + limestone reacts to give Calcium sulphate + Carbon dioxide + water)

Easy! Chemical symbols are like acronyms, a substitute for words. Now it's your turn. Practise with **Try This 24.2.**

$CaCO_3$	Calcium carbonate, limestone
$CaSO_4$	Calcium sulphate
$C_6H_{12}O_6$	One of the monosacharide sugars, glucose or fructose
CH_4	Methane
CO	Carbon monoxide
CO_2	Carbon dioxide
HCl	Hydrochloric acid
H_2O	Water
H_2S	Hydrogen sulphide
H_2SO_4	Sulphuric acid
NOx	Nitrogen oxides, includes NO and NO_2
N_2O	Nitrous oxide
NO	Nitric oxide
NO_2	Nitrogen dioxide
NO^-_2	Nitrite
NO^-_3	Nitrate
NH_3	Ammonia
NH_4	Ammonium
O_3	Ozone
SO_2	Sulphur dioxide
SO_3	Sulphur trioxide

TRY THIS 24.2 – Chemical formulae

Read the following formulae as sentences. What do these relationships describe and what are their geographical significance? Answers on p.269.

$CH_4 + O_2 \rightarrow CO_2 + 2H_2O$

$NH_3 + H_2O \leftrightarrow NH_4 + OH$

$SO_3 + H_2O \rightarrow H_2SO_4$

$12H_2O + 6CO_2 + 709\ \text{kcal} \rightarrow C_6H_{12}O_6 + 6O_2 + 6H_2O$

24.14 An abbreviation and acronym starter list

Most geographers compile a personal acronym and abbreviations list depending on their interests. Add more to make a growing reference resource.

AI	Artificial Intelligence (unless it is an agriculture lecture!)
AOD	Above Ordnance Datum
AONB	Area of Outstanding Natural Beauty
ASCII	American Standard Code for Information Exchange
ASEAN	Association of South East Asian Nations
BATNEEC	Best Available Techniques Not Entailing Excessive Cost
BOD	Biological Oxygen Demand
BPEO	Best Practicable Environmental Option
BTCV	British Trust for Conservation Volunteers
CARICOM	Caribbean Community
CBD	Central Business District
CBR	Crude Birth Rate
CDR	Crude Death Rate
CD-ROM	Compact Disc Read Only Memory
CFCs	Chlorofluorocarbons
CIS	Commonwealth of Independent States
COD	Chemical Oxygen Demand
CPRE	Council for the Protection of Rural England
CSO	Combined Sewer Overflow
CV	Curriculum Vitae
DEM	Digital Elevation Model
DO	Dissolved Oxygen
DTM	Digital Terrain Model
DWF	Dry Weather Flow
EA	Environment Agency, previously NRA – National Rivers Authority and HMIP – Her Majesty's Inspectorate of Pollution
ECLA	Economic Commission for Latin America
EIA	Environmental Impact Assessment
EPA	Environmental Protection Act
ESA	Environmentally Sensitive Area
FAO	Food and Agriculture Organisation, part of United Nations Organisation
FAQ	Frequently Asked Questions
FTP	File Transfer Protocol
FWAG	Farming and Wildlife Advisory Group
GATT	General Agreement On Tariffs And Trade

GIS	Geographic Information Systems
GNP	Gross National Product
GPS	Global Positioning System
HCFCs	Hydrochlorofluorocarbons
HSE	Health and Safety Executive
HTML	HyperText Markup Language
IMF	International Monetary Fund
IPC	Integrated Pollution Control
LDC	Less Developed Country, now usually called Developing Countries
LEAP	Local Environment Agency Plan
LULU	Locally Unacceptable Land Use (nuclear waste, AIDS hospice, see NIMBY)
MAFF	Ministry of Agriculture Fisheries and Food
MBC	Metropolitan Borough Council
MCQs	Multiple Choice Questions
NAFTA	North American Free Trade Association
NATO	North Atlantic Treaty Organisation
NGOs	Non-Government Organisations
NIC	Newly Industrialised Country
NIE	Newly Industrialised Economy
NIMBY	Not In My Back Yard
NIR	Natural Increase Rate
NPP	Net Primary Production
NSA	Nitrate Sensitive Area
NVZ	Nitrate Vulnerable Zone
OAS	Organization of American States
OAU	Organization of African Unity
OECD	Organization for Economic Co-operation and Development
OPEC	Organization of Petroleum Exporting Countries
PCBs	Polychlorinated biphenyls
PCV	Prescribed Concentration Value
pH	Logarithmic scale measuring acidity and alkalinity
ppb	parts per billion
SPSS	Statistical Package for the Social Sciences
SSSI	Site of Special Scientific Interest
TDS	Total Dissolved Solids
TEC	Training and Enterprise Council
TFR	Total Fertility Rate
THES	Times Higher Education Supplement
TLV	Threshold Limiting Value
TNC	Transnational Corporation
TOC	Total Organic Carbon
UNESCO	United Nations Educational, Scientific and Cultural Organization

UNHCR	United Nations Commission on Human Rights
VOCs	Volatile Organic Compounds
WHO	World Health Organization, part of United Nations
WMO	World Meteorological Organization
www	world wide web

Hoehn, P. and Larsgaard, M.L. 1998 *Dictionary of abbreviations and acronyms in geographic information systems, cartography, and remote sensing*, Library, University of California at Berkley, [on-line] http://library.berkeley.edu/EART/abbrev.html Accessed 27 May 1998

24.15 Leaving university

Higher education changes people. Attitudes and values develop and perspectives evolve. It is worth reflecting on how you are altering as a person during your degree course, and think about how that might influence your choice of career. Ask yourself 'How does what I now know about myself and my personal skills and attitudes, influence what I want to do after I graduate?'

Knowing what you want to do when you graduate is difficult and there is an enormous choice available. Your first advice point is your Careers Service. In the absence of a Careers Service do a www search for your local region using careers+service+university. UK students *may* be able to access http://www.prospects.csu.man.ac.uk/STUDENT/CIDD/startpts/casinfo.htm (Accessed 10 January 1999)

Make the most of your geographical skills, the personal, transferable and academic. Look back at Figure 1.1 and check off your skills. Be upfront about them. Tell employers that you have given 20 presentations to groups of 5–50. You have used OHTs, slides and ~~electric~~ electronic display material. Practical skills are easy to list on a CV, but often seem so obvious that they are left out. Familiarity with different database and word processing packages, GIS and mapping packages, modelling, statistics, and hands on laboratory skills are all bonuses of geography degrees. Tell employers you have these skills.

Building and updating a CV throughout your degree course will save time in the last year. There are plenty of texts on CV design, do a key word library search, or check out the Careers Service. There may be an on-line CV designer.

- Always write job applications in formal English, not a casual style. This may seem obvious, but as more applications are sent electronically via www site connections, it is easy to drop into a casual, e-mail writing style, which will not impress human resources managers.

Think of yourself as a marketable product. You possess many skills that employers are seeking, it is a matter of articulating them clearly to maximise your assets.

CARPE DIEM

Ten towns and a river

Add 2 letters in the middle squares to complete the 5-letter towns and river to left and right. When complete, another town will appear. Answers on p.270.

B	R	E			O	K	E
B	A	S			B	A	T
L	E	I			A	M	E
M	A	L			N	Z	A
L	O	I			I	M	S

25 ANSWERS

Geograms 1 p.12

E	C	O	N	O	M	I	C

B	O	U	N	D	A	R	Y

E	U	R	O	P	E	A	N

Try This 2.3 Reflecting on a class or module p.18

A selection of answers from level 2 students:
What I want to get out of attending this module is ... *I want to realise my full communication and research skills/I would like the module to give me a clearer insight into a topic I enjoy and think I might want to pursue as a career/I want to improve my computing skills.*

I have discovered the following about myself with respect to decision-making ... *I now realise that I make more decisions than I realise, but in general I try to avoid the process if possible/I am not particularly decisive, but when I have to make a decision I think things through very carefully/I adopt different decision making processes in different cases/I know it is a very weak point. I think I will find it helpful to understand more about how I make decisions in order to improve. I probably agonise too much/ Sometimes I am really rational and think things through, other times I am totally impulsive.*

What skills did you use well, what skills did other members of the group use well? *As a group we discussed well and no one person took on the role of leader. Everyone listened to everyone else and everyone had valid points to make/I took part in the discussion, however some members of the group did this better.*

What skills were lacking in you (the group) and caused things to go badly? *To start with we had a distinct lack of preparation, hadn't read the briefing material, and generally didn't know where to start/We overestimated the degree of detail required which explains why we took so long/Organising what we were supposed to do and deciding who should do what, wasted some time.*

What did you enjoy least about the exercise/session/module/degree course? *There was a lot of information to evaluate in a short time, a bit of quick thinking required./ Some aspects were boring and appeared unnecessary. Managing time was difficult.*

Wordsearch 1 p.23

Ablate, arroya, Asia, beryl, boreal, bract, canyon, chart, cutting, coral reef, circumpolar, conurbation, eastern, election, empire, foothills, fuel, garden city, Genoa, geyser, hydrate, inner city, Iran, irrigation canal, lagoon, Lima, loess, naze, numeric, observe, rain guage, refuge, region, Reynolds number, runnel, rural, sands, sea, shoal, spinny, suburb, surge, tensor, tertiary, uranic, yardang, zone.

Island links p.32

B	A	L	I		M	U	L	L		L	O	N	G		B	U	T	E
B	A	L	K		M	I	L	L		L	I	N	G		B	A	T	E
B	A	C	K		M	I	L	E		L	I	N	T		F	A	T	E
B	U	C	K		M	I	R	E		L	I	S	T		F	A	R	E
M	U	C	K		E	I	R	E		U	I	S	T		F	A	R	N

Try This 4.2 www addresses p39

All Accessed 10 January 1999.

AGRICOLA http://grc.ntis.gov/agricola.htm or http://www.nal.usda.gov/ag98/
Asian Studies http://coombs.anu.edu.au/WWWVL-AsianStudies.html/
CNN Interactive http://www.cnn.com/
CRIS/USDA http://fundedresearch.cos.com/usda/usda-intro.html
FedWorld Information Network http://www.fedworld.gov/index.html
GeoArchive http://www.lib.utulsa.edu/database/geoarch.htm
GeoRef http://www.agiweb.org/agi/georef/preview.html
Handbook of Latin American Studies http://lcweb2.loc.gov/hlas/
ITAR/TASS http://www.cs.toronto.edu/~mes/russia.html
Reuters http://www.reuters.com/magazine/
Russian Academy of Sciences Bibliographies Indexes http://www.gla.ac.uk/Library/Resources/Databases/ras.html
South China Morning Post http://www.scmp.com/
Wilson Social Science Abstracts http://www.hwwilson.com/socsci.html

Try This 5.3 Where do you read? p.50

Three students' reflections.

'The laundrette is good for skimming, it doesn't matter too much about the interruptions.'
'I have to have peace so we have a house rule about noise after supper for a couple of hours'.
'I get different opportunities to read. I am stuck with a 35-minute bus journey, which is boring but warm! I try to take photocopies of articles on the bus, you can't write notes but you can use a highlighter pen OK-ish. I can use the bus to decide if I need to make notes on something, and what just needs a 5-minute entry and cross ref. in the notes I've got. For real reading I try to make notes at home, but the bus helps because I've usually got some idea about what's going on when I start.'

Try This 5.4 Spotting reading cues p.51

Example phrases.
I shall outline the theory behind ...; The next point is ...; We must first examine and seek to quantify ...; Consider the issues of ...; There is little known about ...; These include ...; Recall ...; Remember ...; However ...; This is the most significant advance ...; We have shown that ...; To summarise the principle points ...; Restating the original argument in the light of this information we see ...; We can conclude that ...; We now know that ...; The important conclusion was ...

Field trip conundrum p.54

Emma and Dave used the augers and got 83%. Jenny and Emily dug soil pits and scored 75%. Eleanor and Gemma measured stream flow and scored 63%. Nigel and Alistair counted pebbles and got 57%.

Lost the plot I p.63

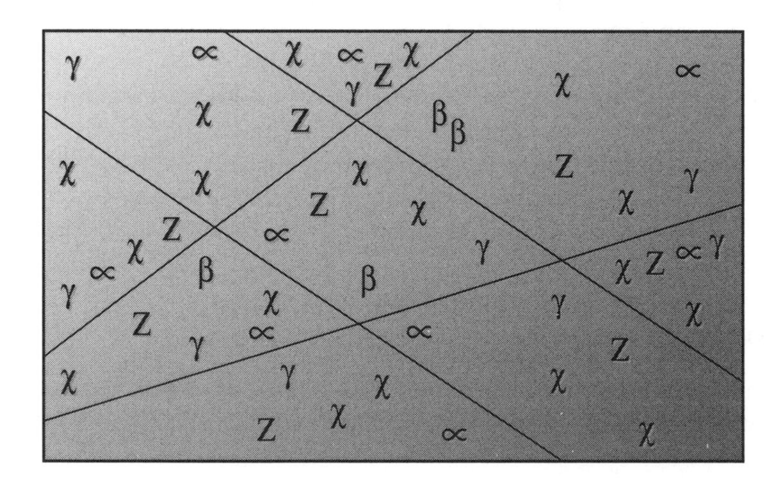

Try This 7.1 Reasoned statements p.69

Some responses, can you do better?

1. *The questionnaire results are right*. We can be confident of the inferences drawn from the questionnaire survey because the sample size was large (678), the criteria established to ensure representative sampling were achieved, and analysis showed statistically significant relationships which confirmed the hypotheses.
2. *People are inflicting potentially catastrophic damage on the atmosphere and causing world-wide climate change*. Information about climate change is currently speculative, but some scientific results support the hypothesis that people are damaging the atmosphere. Smoggit (2010) suggests that increasing concentrations of CO_2 and infra-red absorbing gases released into the atmosphere as waste products of combustion and other human activities, are causing a net rise in global temperatures. Thermals (2010) suggests that one consequence will be higher global temperatures in the next 50 years than previously recorded.
3. *The United States has become an urban country*. At the start of the eighteenth century, 95 per cent of the population of the United States lived in rural communities or on isolated farms (Demoggers 2010). This number declined across two centuries to about 25 per cent, although of the people who live in rural areas as few as 2 per cent are employed in agriculture. The pull of industrialisation and the push of mechanisation in agriculture were major factors in this switch to an urban culture.
4. *Pedestrianisation civilises cities*. Pedestrianisation can create a safer, less polluted and apparently more spacious city. Freeing the centre of cars reduces CO and noise pollution and makes centres more attractive for pedestrians. Shoppers appreciate the calmer atmosphere, and retail-therapy (shopping) is a pleasure rather than a chore (Shoppersareus 2010).

Geo-quick crossword 1 p.74

Across 3 Far flung, 7 Seiche, 8 Iguana, 9 Circle, 10 Nepali, 11 Shop, 13,15 Cliff edge, 17 Nuncio, 18 Isobar, 19 Ararat, 20 Heifer, 21 Entrepot.
Down 1 Zenith, 2 Ice-cap, 3 Federal, 4 Fig Leaf, 5 Unabated, 6 Glaciate, 11 Stone Age, 12 Open cast, 13 Climate, 14 Freight, 15 Exotic, 16 Grazer.

Try This 8.1 Logical arguments? p.81

Some responses.

1. Divergent plate margins occur where two plates are moving away from each other, causing sea-floor spreading. [The first part is a reasonable definition of divergent plate margins. The second part is true where the plate margin is under the ocean. Are there a divergent plate margins on land? There are

examples on Iceland, and what about rift valleys? There is nothing about time scales or the rates involved.]

2. Rainforests store more carbon in their plant tissue than any other vegetation type. Burning forests releases this stored carbon into the atmosphere as CO2. The net result is increased CO_2 in the atmosphere. [Sounds OK, provided the first sentence is factually correct. If it is the *density* of the vegetation in tropical rainforests which maximises the carbon content, then the first sentence is untrue. There are other factors to consider too, like harvesting.]
3. If the system produces a net financial gain then the management regime is successful, and the development economically viable. [This is a rather sweeping statement, no time dimension is mentioned. A net profit might indicate successful management, but it might not be the managers that make the profit. What about maximum profit, could the managers do better? The development maybe successful this year, will this always be the case?]
4. Urban management in the nineteenth century aimed to reduce chaos in the streets from paving, lighting, refuse removal and drainage, and to impose law and order. [The grammar here is awry: is chaos too strong a word? was the paving chaotic? was there any lighting? Did urban managers have wider concerns than sorting out the streets? This is the sort of sentence where you know what the writer is trying to say, but it is slightly off target, general and without precise examples.]

Try This 8.2 Logical linking phrases p.82

Some examples.

and; however; several attempts have been made; conversely; do not agree; if; subsequently; despite the fact; in addition; this has been broadly related to; such data also; these results are opposed to; in other words; coupled with; making a causal relationship; given the pressures to; for instance; one additional mechanism; none the less; such as; how can this; there is evidence; see, for example.

Try This 8.3 Fact or opinion p.82

Some comments in response. The extracts are from articles and reviews in the *Transactions of the Institute of British Geographers*, **22**, 4, 1997.

The city has become 'trendy' (Jencks 1996). p. 411 [Opinion of Jencks, do you need to read Jencks to see if he proved his case? You might need to read the rest of the paper to see what sort of evidence is offered.]

Urban studies has experienced a remarkable renaissance in the past fifteen years, fuelled by the replacement of tight, positivistic approaches with structuralist and, more recently, post-structuralist theories. A veritable deluge of newspaper and magazine reports now addresses urban crises and 'regeneration' processes. p.411 [This is difficult. Is the first sentence the author's conclusion a fact or an opinion that is to be examined? You need to see more of the text to decide. No doubt some urban studies authors would consider the statement a fact, and others disagree. The second sentence is factually correct, but without support as stated here.]

Dam construction has altered flooding patterns in most major river basins in tropical Africa, for example on the Tana River in Kenya or the River Benue in Cameroon downstream of the Lagdo Barrage (Drijver and Marchand, 1985). p.430 [Fact and very nicely exemplified with two case examples and a reference.]

Definitions of 'sustainable development' abound. p.431 [Fact, but unsupported as it stands. This is a clear statement and the author continues with supporting arguments.]

Hansen has described the importance of mountain climbing to the professional middle classes of late nineteenth-century Britain. They 'actively constructed an assertive masculinity to uphold their imagined sense of British imperial power' (Hansen, 1995: 304) p.450 [Plenty of room for debate about this. For middle and upper class British males in the nineteenth century this appeared to be an undoubted fact. Viewed from the perspective of a later century it might seem a little misguided. What do you think?]

Women were unsuited by 'sex and training' for 'exploration' and, since geography was not about library work, like the Asiatic Society, or merely a social club, like the Zoological Society, then, to preserve its focus on exploration, it had to exclude women from its Fellowship (*The Times* 31 May 1893, 11d). p.457 [An opinion held as fact in the 1890s when the RGS was deciding who qualified as Fellows.]

This analysis has documented changes in regulatory practice and the implementation of food policy during a period of increased salience of food quality concerns on the one hand and the intense competitive efforts of corporate retailers to maintain and develop their markets on the other. p.485 [A summary statement that puts the two main thrusts of the paper into one statement. Personally, I find this sentence a little hard to read, could it do with a comma or two? What is 'salience of food quality'? Is this an example of trying too hard to use complex words in writing? BUT beware the temptation to fix on minor criticisms, like the use of a particular word. The main interest is the nature and strength of an argument. Getting side-tracked by liking or loathing an individual's writing style is not helpful.]

The transcriptions of the (40) interviews have been woven into and inform an account of a complex set of (partial and uneven) transitions in the political, cultural and what might best be called the 'moral economy' within which the discipline of geography is reproduced. p.488 [Summary statement that describes the research analysis. This statement appears in the Introduction to the paper, so the author is putting the caveat 'partial and uneven' at the start, so the reader is aware that a synthesis of 40 opinions is not necessarily going to result in an unbiased account.]

Geo-265inks ladder p.88

H	A	Z	E
H	A	L	E
H	O	L	E
H	O	L	D
C	O	L	D
C	O	L	T
C	E	L	T
P	E	L	T
P	E	A	T
M	E	A	T
M	O	A	T
M	O	O	T
M	O	O	R
M	O	O	N
M	O	R	N
T	O	R	N
T	A	R	N
T	A	R	E
D	A	R	E
D	A	N	E
D	U	N	E

Town ladder p.88

L	I	M	A
L	I	M	E
L	I	V	E
H	I	V	E
H	O	V	E
H	A	V	E
H	A	T	E
B	A	T	E
B	A	T	H
B	A	S	H
D	A	S	H
D	I	S	H
D	I	S	S
D	I	M	S
D	I	M	E
D	O	M	E
R	O	M	E
R	I	M	E
R	I	C	E
R	I	C	K
W	I	C	K

Geograms 2 p.89

DOMESTIC CULTURAL STRATEGY

TOPOGRAPHY OXYGENATOR DEMOGRAPHY

MESOZOIC EUROCRAT LATITUDE

POSTMODERN SUBROUTINE EARTHQUAKE

AMMONITE FRONTIER ALLUVIAL

Try This 11.1. Working with a brainstormed list p.105

There are many ways to order this information, the headings here are suggestions. Notice that there is some repetition, a mix of facts and opinions, and some points could go under more than one heading

- Slope instability features: Movement of handrails. Tennis courts on old levelled slip. Evidence that paths across the slips are regularly re-laid, cracked tarmac, fences reset. Promenade held up by an assortment of stilts, walls and cantilever structures where slope below has been undercut. Castle fortification, some walls now undermined and eroded. Path covered by slumping clay. Vegetation eroded from wetter clay slopes.
- Slope repair and restoration features: Concreting the slopes to prevent soil erosion. Fine and coarse netting, geotextiles, used on clay slope to help stabilise the soil. Bolts into the rock face to increase stability. Rock armour installed at the foot of recent slip. Wall to prevent boulders from cliff hitting the road. Tree planting on recent and older slip sites. Drainage holes in retaining walls to reduces soil water pressure.
- Tourism factors: Café for tourists and visitors at foot of slip area. Hotel, amusement arcade and tourist shops – noisy and unsightly. Harbour has mix of marina, fishing and tourist fairground rides. Victorian swimming pool, in need of repair.
- Transport issues: Car parking demands, unsightly in long views. Car pollution. Car parking unscenic. Caravan park on cliff.
- Beach issues: Sea wall prevents undercutting. Rocks at cliff foot dissipate wave energy. Groynes prevent longshore drift. Repairs to sea wall.
- Other issues: Tourist facilities destroy foreshore view. Soil overlying clay rock. Vegetation worn on paths and alongside paths. Litter needs collecting from beach. Road built over old stream, presumed piped underground. Varied quality of coastal path, not always well sign posted.

Try This 11.3 Discussant's role p.110

Offers factual information +	Gives factual information +
Speaks aggressively –	Asks for examples +
Encourages others to speak +	Asks for reactions +
Asks for examples +	Seeks the sympathy vote –
Is very competitive –	Offers opinions +
Helps to summarise the discussion +	Is very defensive -
Keeps quiet –	Summarises and moves discussion to next point +
Ignores a member's contribution –	Gives examples +
Mucks about –	Diverts the discussion to other topics –
Asks for opinions +	Keeps arguing for the same idea,
Is very (aggressively) confrontational –	although the discussion has moved on –

Geo-cryptic crossword 1 p.114

Across 1 Tariffs, 5 Datable, 9 Fecundity, 10 Coypu, 11 Climatologist, 13 Icefield, 15 Linear, 17 Talkie, 19 Diaspora, 22 Rubber stamped, 25 Aorta, 26 Acidifier, 27 Diluted, 28 Mayweed

Down 1 Tufa, 2 Recycle, 3 Fungi, 4 Spiracle, 5 Dry rot, 6 Tectonics, 7 Bayside, 8 Equatorial, 12 Hinterland, 14 Itinerant, 16 Titanium, 18 Liberal, 20 Old Time, 21 Island, 23 Privy, 24 Grid.

Try This 12.1 Keywords in essay questions p.118

1. Asks for 'weaknesses' and 'one' sector only. Don't attempt to impress by covering all *three* sectors *and* strengths *and* weaknesses, which gives an examiner a lot of irrelevant writing to cross out. What is 'contemporary' in this context? Starting with 19th century material will not help your cause.
2. Keywords here are 'growth', 'change' and 'impeded', so there should be little or no explanation of how social structures assist growth and change. Take examples from *either* across Latin America *or* just one country.
3. Wants a technical explanation leading to the derivation of an equation in 1983. Therefore, call on pre-1983 work and the evidence Andrews used, and alternatives and arguments against this particular function. Any reference to post-1983 developments is beyond the scope of this essay. You might refer to subsequent developments in a closing paragraph, but very briefly.
4. A two-part question with equal marks for the 'Account for ...' and 'assess the limits ...'. Evidence and examples are needed for both parts. There is no indication of whether this is a UK, Europe or world-wide question but 'internationalisation' implies a global answer. Beware of answering the first part from UK sector only, if the rest of the answer is globally oriented. You won't lose marks by quoting UK examples, but there are more to be gained from using broader evidence.
5. In 'evaluating' be complementary as well as critical, remember the pros as in raising awareness, as well as the disadvantages. Consider all types of mass media, different types of TV reportage – the news bulletin versus the considered documentary, and the print media, papers and magazines. The answer requires lots of diverse examples. Is this a UK question or should comment be made on media activities in other countries? Incidentally, even if all the lecture examples were from the UK, it does not stop you using European and other examples – unless the module is called something like 'Geography of the UK'.

Try This 12.2 Evaluate an introduction p.120

I think version 2 and 4 are of a good enough standard for a university essay. Versions 1 and 3 contain correct statements, but read more like a bundle of random thoughts than a developed argument. There is much to discuss in this essay so guidelines aid the reader. Version 1 starts promisingly, but the last sentence is unconnected and there is nothing to indicate where the essay is going. Version 3 could introduce a wetlands essay in any country, did you notice Canada was not mentioned? Personally, I do not like the way a definition of wetland is put in brackets in the first sentence of version 3. It is not wrong exactly, but to my mind, important enough not to be in parenthesis. The title of the essay is almost repeated in the third and last sentence. Version 2 is much longer and covers considerable ground! The last sentence indicates that evidence is to be marshalled through case evidence but that all is not bleak, that there is a positive side to mans impact through wetland reclamation and restoration. Version 4 is much slimmer, two main points are made in the first two sentences, and then the essay structure is flagged in the last two. Versions 2 and 4 also benefit from the appropriate embedding of references.

Try This 12.3 Good concluding paragraphs p.124

Version 1 — The tone here is upbeat to the point of tabloidese, 'mega' is not a good adjective and 'explosion' is a little OTT perhaps. The second sentence makes no sense in relation to the first, and ascribes some anthropomorphic attributes to non-sentient wetlands. The fourth sentence does not follow from the third, using the phrase 'This has meant' implies a logical link.

Version 2 — A single sentence paragraph is almost always a bad idea. The sentiments are right but the language very relaxed!

Version 3 — Best of these!

Version 4 — This version starts by sounding more like an introduction and then tails off. Individually the statements are all true, but they don't coalesce.

Version 5 — Starting a final paragraph with 'To conclude', I think, wastes words, the reader can see it is the last paragraph, and the 'This essay has shown' formulae lacks initiative. Not wrong, but not innovative either. The first sentence makes a geographical generalisation. Did people actually have a national policy here, or is the national picture the accumulation of a great many local, independent wetland drainage activities?

Try This 12.6 Shorten these sentences p.129

Wordy	Better
In many cases, the tourists were overcharged.	Many tourists were overcharged.
Microbes are an important factor in soil processes.	Microbes control decay rates in soils.
It is rarely the case that sampling is too detailed.	Sampling is rarely too detailed
The headman was the proud possessor of much of the land in the vicinity of the village.	The headman owned land around the village. OR. The headman was proud to own land near the village.
Moving to another phase of the project	The next phase … or Part x …
The flow rates in the Severn were monitored in August and October respectively.	The flow rates in the Severn were monitored in August and October.
Chi-square is a kind of statistical test. (should read 'Chi-square is a type of statistical test')	Chi-square is a statistical test.
One of the best ways of tackling prison reform is … (this is OK but lacks dynamism)	To tackle prison reform …
The investigation of African cultural impacts continues along the lines outlined.	The investigation of African cultural impacts is continuing.
The nature of the problem …	The problem …
Temperature is increasingly important in snowmelt	Temperature controls snowmelt rates.
One prominent feature of the landscape was the narrow valleys.	Narrow valleys are prominent landscape features
It is sort of understood that …	It is understood that …
It is difficult enough to learn about pollen analysis without time constraints adding to the pressure.	Learning about pollen analysis is time consuming.
The body of evidence is in favour of …	The evidence supports …

Try This 12.8 Synonyms for geoggers? p.131

1 understanding. 2 main. 3 discrimination. 4 extracted. 5 explains. 6 estimates. 7. experience. 8 associated. 9 morphology. (*Could be either, but choose morphology as the more technical term.*) 10 traditional.

Geograms 3 p.134

M	O	B	I	L	I	T	Y

I	N	D	U	S	T	R	Y

F	E	E	D	B	A	C	K

Try This 14.1 The nightmare reference list p.151

This is the same list corrected to be academically acceptable. It is a pain losing essay marks because you do not use the conventions for acknowledging your references. All essays, reports and dissertations must be fully referenced, alphabetically by author, as here:

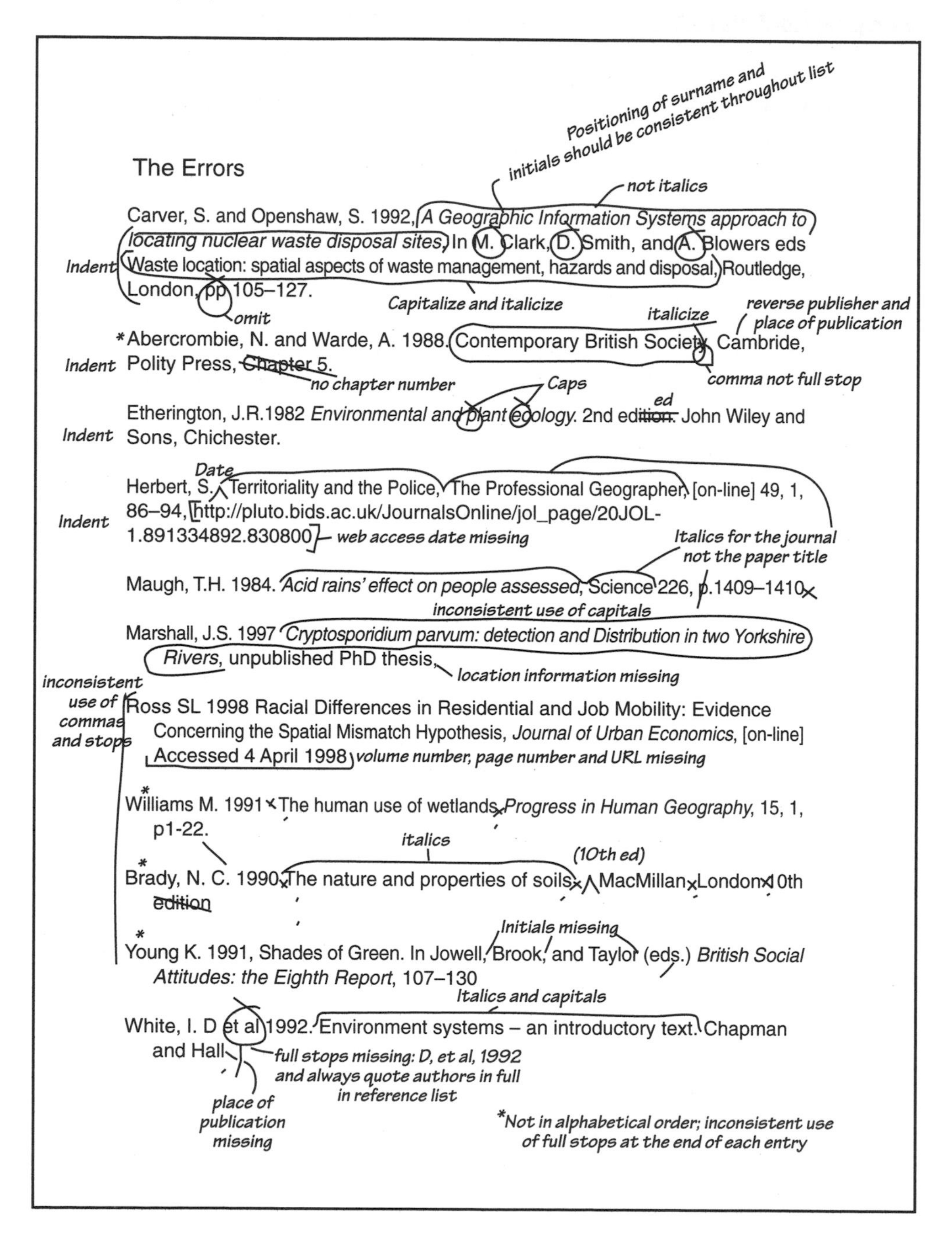

The Errors

Carver, S. and Openshaw, S. 1992, *A Geographic Information Systems approach to locating nuclear waste disposal sites,* In M. Clark, D. Smith, and A. Blowers eds Waste location: spatial aspects of waste management, hazards and disposal, Routledge, London, pp 105–127.

*Abercrombie, N. and Warde, A. 1988. Contemporary British Society. Cambride, Polity Press, Chapter 5.

Etherington, J.R.1982 *Environmental and plant ecology*. 2nd edition. John Wiley and Sons, Chichester.

Herbert, S. Territoriality and the Police, The Professional Geographer, [on-line] 49, 1, 86–94, http://pluto.bids.ac.uk/JournalsOnline/jol_page/20JOL-1.891334892.830800

Maugh, T.H. 1984. *Acid rains' effect on people assessed,* Science 226, p.1409–1410

Marshall, J.S. 1997 *Cryptosporidium parvum: detection and Distribution in two Yorkshire Rivers*, unpublished PhD thesis,

Ross SL 1998 Racial Differences in Residential and Job Mobility: Evidence Concerning the Spatial Mismatch Hypothesis, *Journal of Urban Economics*, [on-line] Accessed 4 April 1998

*Williams M. 1991 The human use of wetlands *Progress in Human Geography*, 15, 1, p1-22.

*Brady, N. C. 1990 The nature and properties of soils MacMillan London 10th edition

*Young K. 1991, Shades of Green. In Jowell, Brook, and Taylor (eds.) *British Social Attitudes: the Eighth Report*, 107–130

White, I. D et al 1992. Environment systems – an introductory text. Chapman and Hall

The corrected list

Abercrombie, N. and Warde, A. 1988 *Contemporary British Society*, Polity Press, Cambridge.

Brady, N.C. 1990 *The Nature and Properties of Soils*, (10th edn), MacMillan, London.

Carver, S. and Openshaw, S. 1992 A Geographic Information Systems Approach to Locating Nuclear Waste Disposal Sites. In Clark, M., Smith, D. and Blowers, A. (eds) *Waste Location: Spatial Aspects of Waste Management, Hazards and Disposal*, Routledge, London, 105–127.

Etherington, J.R. 1982 *Environmental and Plant Ecology* (2nd edn), John Wiley and Sons, Chichester.

Herbert, S. 1997 Territoriality and the Police, *The Professional Geographer*, [on-line] 49, 1, 86–94, http://pluto.bids.ac.uk/JournalsOnline/jol_page/20JOL-1.891334892.830800 Accessed 5 April 1998.*

Marshall, J.S. 1997 *Cryptosporidium parvum:* Detection and Distribution in Two Yorkshire Rivers, unpublished PhD thesis, University of Leeds.

Maugh, T. H. 1984 Acid rain's effect on people assessed, *Science*, 226, 1408–1410.

Ross, S.L. 1998 Racial Differences in Residential and Job Mobility: Evidence Concerning the Spatial Mismatch Hypothesis, *Journal of Urban Economics*, [on-line] 43, 1, 112–135, http://pluto.bids.ac.uk/JournalsOnline/jol_page/20JOL-1.891334297.488900 Accessed 4 April 1998.*

White, I.D., Mottershead, D.N. and Harrison, S.J. 1992 *Environmental Systems: an introductory text*, Chapman and Hall, London.

Williams, M. 1991 The Human Use of Wetlands, *Progress in Human Geography*, 15, 1, 1–22.

Young, K. 1991 Shades of Green. In Jowell, R., Brook, L. and Taylor, B. (eds) *British Social Attitudes: the Eighth Report*, 107–130.

* n.b. You cannot access the Herbert or Ross reference without a password, hence the code numbers after JOL.

Where are we? Wordsearch 2 p.153

Ankara, Armenia, Azov, Belarus, Black, Bucharest, Bulgaria, Chisinau, Danube, Dnepr, Don, Estonia, Finland, Helsinki, Kiev, Latvia, Lithuania, Minsk, Moldovia, Moscow, Poland, Riga, Rumania, Sweden, St. Petersburg, Stockholm, Tallinn, Tblisi, Turkey, Ukraine, Vilnius, Vistula, Volga, Warsaw, White, Yerevan.

Try This 16.1 Book reviews p.160

Comments on the extracts from Book Reviews published in *Area*, **29**, 3, 1997 (page numbers indicated).

1. The appearance of *Disability and the City* represents an important moment for the discipline, being the first major geographic analysis of disability in published form. (p.274). [A very supportive statement explaining that this is a landmark publication.]

2. However, it is the scope and scale of the empirical material presented in this volume that mark its greatest achievement. Questions of gender and race, restructuring, spatiality, consumption and politics are explored in 15 different chapters, each providing a window onto theoretical abstractions. (p.276) [The first sentence is supportive and positive. The second lists the key areas covered and indicates the scale of the text, 15 chapters, and indicates the establishment of links between empirical and theoretical material.]
3. *Glacial Geology: Ice Sheets and Landforms* is very up-to-date and well written. It covers an extremely wide range of topics and concludes with an interesting chapter on 'Interpreting Glacial Landscapes'. It is written for students and I expect that they will really like it. (p 279) [Clearly this reviewer is happy that this is a well produced and academically up to the minute production that will be accessible to undergraduates. The reviewer is indicating that the authors have achieved their intention of writing for the undergraduate market.]
4. Because of the breadth of subject matter, certain individual sections are quite sparse. Also, I feel that in places there is a strange combination of new and old ideas presented side by side, even though they may be contradictory, e.g. there is much discussion of lodgement tills and melt-out tills but no suggestion of the debate over their distinguishing criterion and large-scale existence. (p.279) [Clearly the first sentence is a reservation, but with the caveat that because this is a general text there cannot be in-depth treatment of all aspects. The second sentence is another reservation and very nicely exemplified. An example of a supported argument.]
5. It is important therefore that any introductory text concisely defines its subject matter at the outset. Nowhere, however, do we gain a clear indication of what environmental science *is* (or is *not* for that matter). (p.280) [An unhappy reviewer, pointing out what is, for him, a fundamental flaw.]
6. The book sets out to provide a manual of landslide recognition for the non-specialist and it achieves this objective admirably. Its focus on European features might mean that some of the most spectacular world examples are not included, but it also means that there are landslides which I am sure have not previously appeared in the English literature. (p.281) [Positive indication that the book fulfils its mission and not just in a predictable manner. Even if you know all about landslides you might want to chase up this text to look at the pictures.]
7. It has taken three years for this selection of papers to see the light of day. … One might question the purpose of having such delayed, if thoroughly refereed, proceedings, when many of the results have been overtaken by events elsewhere, not least by the contributing authors. (p. 282) [A fairly damming statement, that will not increase the sales. It points up a perennial problem with books. Preparation and publication timescales can push contents out of date very quickly. Hence the emphasis on recent journal publications on reading lists.]
8. In the time between the San Diego meeting and the appearance of the book,

other advanced texts showing the forefront of research with a better coverage have appeared, for example *Lancaster's Geomorphology of Desert Dunes* (1995) and Abrahams and Parsons' *Geomorphology of Desert Environments* (1993). (p. 283) [Picks up the point about delayed publication which in this case is more obvious because the reviewer can quote two texts published in the field in the intervening period. If this was the only book on the topic and there were no rivals in sight the reviewer might be less hard on the authors.]

9. These chapters are rather more variable in quality and some deserve scant attention in the opinion of the reviewer. Is it really of any cartographic value, for example, to use sketch maps produced by 124 undergraduates to postulate as to gender differences in map reading abilities (Kumler and Butterfield, Chapter 10)? How, one wonders, is the fact that apparently 32 per cent of males and 14 per cent of females put north at the top of their maps, going to add to our knowledge of map design? (p.284) [The author clearly has some reservations about some of the chapters and illustrates his point with an example. Do you feel you need to read the book to find out if the reviewer is being fair? Are Kumler and Butterfield being taken out of context? Is this a typical example of all the work in these chapters, or the most extreme example that can be quoted? You would need to go back to the book. You might also look to see who is writing the review. Is this someone whose opinion is to be respected? Clearly the editors of *Area* thought so, and have published this review, which is one criteria of quality. It is generally good advice to be wary of either over effusive endorsements or very extreme criticism when reading or writing reviews.]
10. The writers ought to be congratulated on producing such a clear and informative text covering a broad range of issues within the field of people and the environment. It is rare for an edited volume to have the coherency contained in this text. (p.285) [Time to buy, and good value at £11.95]

Geo-quick crossword 2 p.181

Across 1 Census, 5 Arch, 8 Etna, 9 Panorama, 10 Integral, 11 Lath, 12 Safari, 14 Newton, 16 Mere, 18 Humidity, 20 Artifact, 21 Dail, 22 Bray, 23 Resort
Down 2 Estonia, 3 State, 4 Superhighway, 5 Airflow, 6 Comet, 7 Inclinometer, 13 Amenity, 15 Outlier, 17 Error, 19 Dodos.

Try This 19.5 Generic questions for revising geography! p.189

1. What is the purpose of ...? (*cryosphere, vector GIS model, Lexis diagram, public–private partnerships*)
2. Why is ... an inadequate explanation of ...?
3. Who are the three main authors to quote for this topic?
4. Name two examples not in the course text or lectures for ...?
5. Explain the findings of ...?
6. What are the main characteristics of ...?

7. Outline the relationship between ... and ...
8. What are the limitations of the ... (methodological) approach? (*environmental impact assessment, core-periphery perspective, neural network*)
9. Define ...?/What does ... mean? (*aquaculture, hegemony, ecotaxation, cultural ecology, porosity*)
10. What are the limitations of ...? (*holistic land management? north-south divide concept, ice-cores as climate records*)
11. What methodology is employed to ...?
12. At what point does this process become a hazard?/of concern? (*algal blooms, monopolies*)
13. This is an unusual result. Why is this so?
14. How has human impact affected ...?
15. What is the purpose of ...? (*management, social analysis, hazard mapping*)
16. What was the main aim of ...? (*open-ended questions?, reconstructing past climates*)
17. What is the spatial scale involved here?
18. How is ... calculated? (*Population density, Reynolds number, market potential*)
19. Outline the sequence of events that involved in ...? (*El Nino, urban regeneration, terracing*)
20. If ... did not occur, what would be the implications? (*fire in ecosystems, inward investment*)
21. How has ... adapted to ...? (*How has vegetation adapted in arid environments?, Housing tenure changed in post-war Britain?*)
22. Outline the different approaches that can be taken to the study of ... (*ecosystems, ethnic diversity*)
23. Will this work in the same way at a larger / smaller scale?
24. How influential has ... been? (*central place theory*)
25. What are the ... (*regional, national, cultural*) ... implications of this finding?
26. What is meant by ...? (*commensalism, niche, logical positivism, crude death rate, safety factor*)
27. Why is ... important for ...? (*copper ... for plant growth , secondary census data ... for population studies*)
28. Do I agree that ...? ... (*the nineteenth century saw a golden era of initiative and independence in local government*)
29. Outline three types of ...?

Geograms 4 p.190

A	N	T	A	R	C	T	I	C	A

M	A	A	S	T	R	I	C	H	T

P	L	A	N	T	A	T	I	O	N

W	I	L	D	E	R	N	E	S	S

M	E	T	R	O	P	O	L	I	S

E	S	C	A	R	P	M	E	N	T

Geographical links p.199 Regional

Lost the plot 2 p.204

Geojumble p.219

The upper diagram has a skate board, wellington boot and pear. The lower diagram has a rucksack, trainer, apple and beer pull.

Geo-cryptic crossword 2 p.227

Across 1 Landslip, 6 Orient, 9 Ultrabasic, 10 Fens, 11 Chalcopyrite, 13 Chou, 14 Lee waves, 17 Flat spin, 18 Gite, 20 Nomenclature, 23 Zebu, 24 Bridgeable, 25 Raster, 26 Medieval

Down 2 Able, 3 Duricrust, 4 Libyan, 5 Postcolonialism, 6 Occupier, 7 Infer, 8 Nineteenth, 12 Phylloxera, 15 Aggregate, 16 Spacebar, 19 Rugged, 21 Erupt, 22 Alga.

Try This 24.1 The summer of 1998 quiz p.230

1j, 2p, 3k, 4m, 5q, 6o, 7d, 8t, 9s, 10l, 11g, 12h, 13b, 14i , 15r, 16c, 17e, 18f, 19n, 20a.

Try This 24.2 Chemical formulae p.247

$CH_4 + O_2 \rightarrow CO_2 + 2H_2O$

When methane and oxygen are burned together the reaction gives carbon dioxide and water, a greenhouse gas issue.

$NH_3 + H_2O \leftrightarrow NH_4 + OH$

Describes the ammonia to ammonium oxidation – reduction interaction, a soils process.

$SO_3 + H_2O \rightarrow H_2SO_4$

An atmosphere interaction where sulphur trioxide and water combine to form sulphuric acid, which acidifies rain.

$12H_22O + 6CO_2 + 709\ kcal \rightarrow C_6H_{12}O6 + 6O_2 + 6H_2O$

Describes the synthesis of water and carbon dioxide with light energy in plant leaves to produce the carbohydrate $C_6H_{12}O_6$ for plant growth with water and oxygen as waste products, – a biogeography issue.

Ten towns and a river p.251 Strathmore.

26 STUDY SKILLS: A MINI-BIBLIOGRAHY

This is a mini-bibliography of skills texts. A compilation of the references from each chapter is at http://www.geog.leeds.ac.uk/staff/p.kneale/skillbook.html

Barnes, R. 1995 *Successful Study for Degrees*, (2nd edn) Routledge, London

Bucknall, K. 1996 *Studying at University: how to make a success of your academic course*, How To Books, Plymouth

Buzan, T. 1989 *Speed Reading*, David and Charles, Newton Abbott

Drew, S. and Bingham, R. (eds) 1997 *The Student Skills Guide*, Gower, Aldershot

Fairbairn, G.J. and Winch, C. 1996 *Reading, Writing and Reasoning: a guide for students*, (2nd ed.) Open University Press, Milton Keynes

Freeman, R. 1991 *Mastering Study Skills*, MacMillan Press, London

Hargie, O., Saunders, C. and Dickson, D. 1987 *Social Skills in Interpersonal Communication*, (2nd edn) Croom Helm, London

Hay, I., Bochner, D. and Dungey, C. 1997 *Making the Grade: a guide to successful communication and study*, Oxford University Press, Australia.

Hector-Taylor, M. and Bonsall, M. 1994 *Successful Study: a practical way to get a good degree*, (2nd edn) The Hallamshire Press, Sheffield

Hind, D.W.G. 1989 *Transferable Personal Skills – a student guide*, Business Education Publishers, Sunderland

Kirkman, J. 1993 *Full Marks: advice on punctuation for scientific and technical writing*, (2nd edn) Ramsbury Books, Malborough

Maddox, H. 1988 *How to Study*, (Rev. edn), Pan Books, London

Marshall, L.A. and Rowland, F. 1983 *A Guide to Learning Independently*, Open University Press, Milton Keynes

Newby, M. 1989 *Writing: a guide for students*, Cambridge University Press, Cambridge

Northedge, A. 1990 *The Good Study Guide*, The Open University, Milton Keynes

Northedge, A., Thomas, J., Lane, A. and Peasgood, A. 1997 *The Sciences Good Study Guide*, The Open University, Milton Keynes

Rose, C. 1985 *Accelerated Learning*, Accelerated Learning Systems, Great Missenden, UK

Rowntree, D. 1993 *Learn How to Study: a guide for students of all ages*, (3rd edn), Warner Books, London

Rudd, S. 1989 *Time Manage Your Reading*, Gower, Aldershot

Russell, S. 1993 *Grammar, Structure and Style*, Oxford University Press, Oxford

Saunders, D. (ed.) 1994 *The Complete Student Handbook*, Blackwell, Oxford
Selmes, I. 1987 *Improving Study Skills*, Hodder and Stoughton, London
Van den Brink-Budgen, R. 1996 *Critical Thinking for Students: how to use your recommended texts on a college or university course*, How To Books, Plymouth

What do you call a geomorphologist who repairs his own equipment?

INDEX